A BIOLOGIST'S ADVANCED MATHEMATICS

DAVID R. CAUSTON

Department of Botany and Microbiology,
University College of Wales,
Aberystwyth

London
ALLEN & UNWIN
Boston   Sydney   Wellington

Allen & Unwin, the academic imprint of

Unwin Hyman Ltd

PO Box 18, Park Lane, Hemel Hempstead, Herts HP2 4TE, UK
40 Museum Street, London WC1A 1LU, UK
37/39 Queen Elizabeth Street, London SE1 2QB

Allen & Unwin Inc.,
8 Winchester Place, Winchester, Mass. 01890, USA

Allen & Unwin (Australia) Ltd,
8 Napier Street, North Sydney, NSW 2060, Australia

Allen & Unwin (New Zealand) Ltd in association with the Port Nicholson
Press Ltd,
60 Cambridge Terrace, Wellington, New Zealand

First published in 1987

British Library Cataloguing in Publication Data

Causton, David R.
    A biologist's advanced mathematics.
1. Biomathematics
I. Title
510′.24574   QH323.5
ISBN 0-04-574036-4

Library of Congress Cataloging-in-Publication Data

Causton, David, R.
    A biologist's advanced mathematics.
Rev. ed. of: A biologist's mathematics, 1977.
Bibliography: p.
Includes index.
1. Biomathematics. I. Causton, David, R. Biologist's
mathematics. II. Title
QH323.5.C38   1987   510′.24574   86-20641
ISBN 0-04-574036-4 (alk. paper)

Set in 10 on 12 point Times by Mathematical Composition Setters Ltd,
Salisbury, Wiltshire and printed in Great Britain

Preface

This book is the second of two which are effectively the second edition of my *A biologist's mathematics*, first published in 1977. The first of the current two books, *A biologist's basic mathematics*, was published by Edward Arnold in 1983 and constituted the more elementary two-thirds or so of the former book with only relatively minor changes. The present book, although incorporating the remainder of the original, is essentially a new work designed for advanced undergraduates, postgraduates, and research workers in the biosciences and agrosciences. The main purpose of the book is to ease the transition from elementary mathematical knowledge to the level required for reading books on advanced special topics, such as E. C. Pielou's *Mathematical ecology*, J. H. M. Thornley's *Mathematical models plant physiology*, D. S. Jones and B. D. Sleeman's *Differential equations and mathematical biology*, M. Bazin's *Mathematics in microbiology*, D. R. Causton and J. C. Venus's *The biometry of plant growth*, and J. France and J. H. M. Thornley's *Mathematical models in agriculture*.

It should be clearly understood that the term 'advanced' in the title refers to the level of mathematical difficulty from the viewpoint of the average biologist; in no sense does the book deal with advanced mathematics as such. The readers for whom the book is designed are more likely to select topics they require from time to time rather than to be pursuing a systematic course, and so the ordering of chapter topics is somewhat arbitrary but is broadly within a framework of increasing mathematical difficulty. Consequently there are some instances of the use of topics earlier in the book which are not introduced and treated in detail until later on; such cases are adequately page referenced.

The selection of mathematical topics for a book of this nature presents some difficulty. Although there are several main themes that are obviously essential, for example linear algebra and differential equations, the choice of some of the precise detail within such broad subject areas is not always so clear cut. No attempt has been made to have a comprehensive coverage of everything possible that any biologist might require in the area of mathematics, but it is hoped that the needs of the majority will be served most of the time. Unlike in *A biologist's basic mathematics*, for the present book a knowledge of elementary univariate statistical theory and methods is assumed, such as is given to nearly all biological and agricultural students nowadays. This considerably increases the scope for discussing useful applications of what in the first instance are purely mathematical topics.

The philosophy underlying the presentation of applications also requires some explanation. In one sense, a book at this level does not need biological illustrative material applying the mathematics discussed, since readers will either: (a) already have applications in mind for which they require the mathematical assistance that I attempt to provide in this book; or (b) be wanting to read a book of the kind indicated above, which is mostly or wholly concerned with applications. However, it is felt that biological applications of mathematics should be included whenever, for diverse reasons, it seems pertinent to do so. Thus, for example, Principal Component Analysis is discussed at great length in Chapter 2 because I feel that there is no really satisfactory account of the biological interpretation of the analysis available elsewhere. Similarly, several examples of the application of differential equations are given, both because of their importance and because of their ease of presentation. However, the result of this is a patchiness of intensity of applications throughout the book as a whole, with large sections being devoid of any direct discussion of biological usage. But it does seem pointless to reproduce a biological application simply to preserve a balance in this respect when a good account is available from some other source.

Most of the chapters in this book start where *A biologist's basic mathematics* stops, and readers are assumed to have a mathematical knowledge of equivalent level to that book. The aim in Chapter 1 is to continue matrix theory, covering determinants and inversion, then to give a systematic introduction to linear equations, and finally to return to matrices to study eigenvalues and eigenvectors, linear dependence of vectors, and matrix rank. Much of the application of linear algebra, in Chapter 2, is concerned with multivariate statistics, together with more on population growth and some direct biological applications of linear equations.

Chapter 3 covers many topics in the integral calculus: methods of integration, infinite integrals, gamma and beta functions which are defined by integrals, and multiple integrals. Most of this material is then applied to the mathematics of statistical probability densities and a model of tree stem shape. Although placed in this chapter, the section on multiple integrals should not be read until much of Chapter 6 has been assimilated.

Trigonometric functions are dealt with in Chapter 4, together with inverse trigonometric functions. There is also a note on hyperbolic functions and an introduction to Fourier series. Chapter 5 is a mixture of no fewer than four unconnected main topics, none of which require a whole chapter to themselves. These are series – particularly the Maclaurin and Taylor

iv

series and their applications; complex numbers; a systematic study of partial fractions; and a short, largely descriptive, section on numerical analysis.

The purpose of Chapter 6 is to introduce the mathematics of functions of more than one variable through a consideration of selected functions of two variables and their associated geometrical representation. This introduces such ideas as direction cosines and saddle points. The surfaces of the functions discussed are shown in three-dimensional graphs, and the two-dimensional contour curves method of graphing the functions is also presented. Partial differentiation and allied topics are covered at the end of the chapter.

Although the fitting of mathematical functions to data is basically a statistical subject, Chapter 7 is devoted to a consideration of function fitting by the method of least squares. By confining attention to function fitting *per se,* and by omitting details of setting standard errors of parameter estimates and of the various tests of hypotheses which one can make about the fitted curve, a largely non-statistical account is given. Nevertheless, both linear and non-linear regression situations are discussed, including some of the difficulties encountered with the latter.

Differential equations are of central importance in many areas of biology, and the final two chapters of the book are devoted to them. Although much of the material of these two chapters is concerned with analytical methods of solution of equations of different types, the philosophy at the outset is to recognise that most differential equations arise in practice in the formulation of mathematical models of complex physical and biological phenomena and are not usually amenable to analytical solution. Thus Chapter 8 starts with a general discussion of mathematical models, and then the properties of several simple equations and their numerical, as well as their analytical, solution are considered. The remainder of the chapter is devoted to linear differential equations of both first and second order, employing both the classical methods and the Laplace transform method of solution. The subject of linear differential equations is continued in the early part of Chapter 9, in which simultaneous differential equations are discussed. This final chapter is continued with a brief consideration of certain non-linear equations that are amenable to analytical solution, and finally there is an introduction to stability analysis, which, relying as it does on many mathematical topics already discussed, is conveniently placed at the end of the book.

Most chapters have a list of suggestions for further reading and/or references to specific applications of the mathematical topics discussed in the chapter. Similarly, some unworked examples as exercises appear at the end of most chapters.

I am very grateful to Dr Geoffrey M. Clarke and Dr John Grace for reading through the typescript (excepting Chapters 7 and 9) and making valuable comments and suggestions. Also, my thanks go to Professor W. G. Hill, Professor P. H. A. Sneath, Dr M. B. Usher, and Dr John Grace for recommendations of topics for addition to (and also suggestions of deletions from!) my original plan and a partially written early version. Inevitably, however, I have not always heeded advice; so, as for any author, the responsibility for the material finally presented in this book is my own. Last, but certainly not least, I am indebted to my friend Mr Miles Jackson, formerly of Allen & Unwin, for his unceasing encouragement and helpfulness.

D. R. CAUSTON

Contents

CONTENTS

List of tables

1

Linear algebra

A glance through the pages of this chapter might seem to suggest to the reader that the subject matter is essentially matrix algebra. This is partially true although, unlike the final chapter in Causton (1983), the presentation of more matrix algebra methodology is not the sole object of this chapter. An important topic in mathematics and its applications is that of linear equations. You will already have been introduced to this topic at school under the name of 'simultaneous equations' using two or three equations containing, respectively, two or three unknowns. However, in practice, not only can we have a large set of equations in many unknowns but the number of equations and unknowns may be unequal; and even if they are equal the equations many not be solvable in a straightforward manner. Matrix algebra is a cornerstone in the subject of linear equations, although the latter is by no means the only application of the former; and other applications of matrix algebra will also be introduced in the next chapter. Nevertheless, the twin subjects of linear equations and matrix algebra are so intertwined that they often appear together under the name of **linear algebra**.

The idea of the **inverse of a matrix** was introduced in Causton (1983). For a matrix $\mathbf{A}$, the inverse $\mathbf{A}^{-1}$ has the property that

$$\mathbf{A}\mathbf{A}^{-1} = \mathbf{I}$$

or, equally well,

$$\mathbf{A}^{-1}\mathbf{A} = \mathbf{I}$$

where $\mathbf{I}$ is the unit matrix. We now require a method for finding the inverse matrix; in presenting such a method another topic in matrix algebra will also be introduced.

Determinants

The inverse, and determinant, of a second-order matrix

It has already been stated that the inverse, $\mathbf{A}^{-1}$, of any square matrix $\mathbf{A}$ is

1

such that $\mathbf{AA}^{-1} = \mathbf{I}$. Consider the second-order matrix

$$\begin{bmatrix} a & c \\ b & d \end{bmatrix}$$

where, for clarity, we use a separate letter for each element rather than the usual subscript notation.

Now, form a new matrix from the above by reversing the two elements in the leading diagonal and changing the signs of the other two, i.e.

$$\begin{bmatrix} d & -c \\ -b & a \end{bmatrix}$$

and consider the product of these two matrices:

$$\begin{bmatrix} a & c \\ b & d \end{bmatrix} \begin{bmatrix} d & -c \\ -b & a \end{bmatrix} = \begin{bmatrix} ad - bc & -ac + ac \\ bd - bd & -bc + ad \end{bmatrix} = \begin{bmatrix} ad - bc & 0 \\ 0 & ad - bc \end{bmatrix}$$

$$= (ad - bc) \begin{bmatrix} 1 & 0 \\ 0 & 1 \end{bmatrix} = (ad - bc)\mathbf{I}$$

The product of these two matrices yields the unit matrix multiplied by the scalar quantity $(ad - bc)$. Now consider the following matrix multiplication:

$$\begin{bmatrix} a & c \\ \\ b & d \end{bmatrix} \begin{bmatrix} \dfrac{d}{ad - bc} & \dfrac{-c}{ad - bc} \\ \\ \dfrac{-b}{ad - bc} & \dfrac{a}{ad - bc} \end{bmatrix} = \begin{bmatrix} \dfrac{ad}{ad - bc} + \dfrac{-bc}{ad - bc} & \dfrac{-ac}{ad - bc} + \dfrac{ac}{ad - bc} \\ \\ \dfrac{bd}{ad - bc} + \dfrac{-bd}{ad - bc} & \dfrac{-bc}{ad - bc} + \dfrac{ad}{ad - bc} \end{bmatrix}$$

$$= \begin{bmatrix} \dfrac{ad - bc}{ad - bc} & 0 \\ \\ 0 & \dfrac{ad - bc}{ad - bc} \end{bmatrix} = \begin{bmatrix} 1 & 0 \\ 0 & 1 \end{bmatrix}$$

This time the product is the unit matrix, and since the matrix

$$\begin{bmatrix} \dfrac{d}{ad - bc} & \dfrac{-c}{ad - bc} \\ \\ \dfrac{-b}{ad - bc} & \dfrac{a}{ad - bc} \end{bmatrix}$$ has been formed entirely from $\begin{bmatrix} a & c \\ b & d \end{bmatrix}$

the former matrix must be the inverse of the latter. The quantity $(ad - bc)$ is called the **determinant** of the matrix and, as already mentioned, it is a scalar quantity.

The rule for forming the inverse of a second-order matrix can therefore be put into words as follows.

2

The inverse of a second-order matrix is formed by reversing the elements in the leading diagonal, changing the signs of the remaining two elements, and then dividing each of the resulting new elements by the determinant of the matrix.

Example 1.1   Find the inverse of

$$\begin{bmatrix} 4 & -6 \\ 2 & 5 \end{bmatrix}$$

and verify the result.

The determinant is $(4)(5) - (2)(-6) = 20 - (-12) = 32$. Thus the inverse is

$$\frac{1}{32}\begin{bmatrix} 5 & 6 \\ -2 & 4 \end{bmatrix} = \begin{bmatrix} \frac{5}{32} & \frac{3}{16} \\ -\frac{1}{16} & \frac{1}{8} \end{bmatrix}$$

Verifying:

$$\frac{1}{32}\begin{bmatrix} 4 & -6 \\ 2 & 5 \end{bmatrix}\begin{bmatrix} 5 & 6 \\ -2 & 4 \end{bmatrix} = \frac{1}{32}\begin{bmatrix} 20+12 & 24-24 \\ 10-10 & 12+20 \end{bmatrix}$$

$$= \frac{1}{32}\begin{bmatrix} 32 & 0 \\ 0 & 32 \end{bmatrix} = \begin{bmatrix} 1 & 0 \\ 0 & 1 \end{bmatrix}$$

Determinants arose independently of, and long before, matrix algebra. In order to introduce the notation of the subject, consider the matrix

$$\mathbf{A} = \begin{bmatrix} a & c \\ b & d \end{bmatrix}$$

The determinant of **A** can be symbolised in various ways:

$$|A| = \det \mathbf{A} = \Delta = \begin{vmatrix} a & c \\ b & d \end{vmatrix} = ad - bc \tag{1.1}$$

The first symbol in (1.1) shows the letter denoting the matrix not in heavy type but enclosed by vertical lines. The second symbol is obvious, but the third symbol is ambiguous since it can refer to the determinant of any matrix. The fourth symbol looks like the original matrix, but vertical lines replace the square brackets; and finally in (1.1) is the definition of the determinant.

Third- and higher-order determinants

The idea of a determinant can be extended to higher orders. Consider the

3

3×3 array

$$\mathbf{A} = \begin{bmatrix} a & d & g \\ b & e & h \\ c & f & i \end{bmatrix}$$

where again for clarity the subscript notation is not used. What meaning is to be ascribed to $|A|$?

For any particular element in

$$\begin{vmatrix} a & d & g \\ b & e & h \\ c & f & i \end{vmatrix}$$

all other elements *not* in the same row or column as the element itself form a 2×2 determinant, called a **minor**. Thus, for the determinant in hand, we have the situation shown in the first two columns of Table 1.1. Moreover, if each minor is given a positive or negative sign according to the scheme shown in the third column of the table, then the resulting 'signed minors' are called **co-factors** and are shown in the last column of Table 1.1.

Now select any row or column of $|A|$, multiply each of the three elements in that row or column by their corresponding co-factors, and add the three products together: the result is $|A|$. If we select the first column of the determinant, then we have

$$\begin{vmatrix} a & d & g \\ b & e & h \\ c & f & i \end{vmatrix} = a \begin{vmatrix} e & h \\ f & i \end{vmatrix} + b \left\{ - \begin{vmatrix} d & g \\ f & i \end{vmatrix} \right\} + c \begin{vmatrix} d & g \\ e & h \end{vmatrix}$$

$$= a(ei - fh) - b(di - fg) + c(dh - eg)$$

$$= aei - afh - bdi + bfg + cdh - ceg$$

We could equally well have selected another row or column; thus, employing the middle row:

$$|A| = b \left\{ - \begin{vmatrix} d & g \\ f & i \end{vmatrix} \right\} + e \begin{vmatrix} a & g \\ c & i \end{vmatrix} + h \left\{ - \begin{vmatrix} a & d \\ c & f \end{vmatrix} \right\}$$

$$= -b(di - fg) + e(ai - cg) - h(af - cd)$$

$$= -bdi + bfg + aei - ceg - afh + cdh$$

If this expression for the determinant is compared with the previous one, they will be found to be identical.

Table 1.1 Details of minors and co-factors in a third-order determinant.

Element	Minor	Pattern of signs for co-factor	Co-factor
a	$\begin{vmatrix} e & h \\ f & i \end{vmatrix}$		$\begin{vmatrix} e & h \\ f & i \end{vmatrix}$
b	$\begin{vmatrix} d & g \\ f & i \end{vmatrix}$		$-\begin{vmatrix} d & g \\ f & i \end{vmatrix}$
c	$\begin{vmatrix} d & g \\ e & h \end{vmatrix}$		$\begin{vmatrix} d & g \\ e & h \end{vmatrix}$
d	$\begin{vmatrix} b & h \\ c & i \end{vmatrix}$		$-\begin{vmatrix} b & h \\ c & i \end{vmatrix}$
e	$\begin{vmatrix} a & g \\ c & i \end{vmatrix}$	$\begin{matrix} + & - & + \\ - & + & - \\ + & - & + \end{matrix}$	$\begin{vmatrix} a & g \\ c & i \end{vmatrix}$
f	$\begin{vmatrix} a & g \\ b & h \end{vmatrix}$		$-\begin{vmatrix} a & g \\ b & h \end{vmatrix}$
g	$\begin{vmatrix} b & e \\ c & f \end{vmatrix}$		$\begin{vmatrix} b & e \\ c & f \end{vmatrix}$
h	$\begin{vmatrix} a & d \\ c & f \end{vmatrix}$		$-\begin{vmatrix} a & d \\ c & f \end{vmatrix}$
i	$\begin{vmatrix} a & d \\ b & e \end{vmatrix}$		$\begin{vmatrix} a & d \\ b & e \end{vmatrix}$

Example 1.2 Evaluate

$$\begin{vmatrix} -1 & 2 & -3 \\ 2 & -1 & 4 \\ 3 & 4 & 1 \end{vmatrix}$$

Utilising the first column, we have

$$= -1\begin{vmatrix} -1 & 4 \\ 4 & 1 \end{vmatrix} + 2\left\{-\begin{vmatrix} 2 & -3 \\ 4 & 1 \end{vmatrix}\right\} + 3\begin{vmatrix} 2 & -3 \\ -1 & 4 \end{vmatrix}$$

$$= -1(-1 - 16) - 2(2 + 12) + 3(8 - 3) = 17 - 28 + 15 = 4$$

Determinants of higher-order matrices are evaluated in an analagous fashion. For each element there is a corresponding co-factor, and the determinant is calculated by multiplying all the elements in one row or column by their co-factors and summing the resulting products. The signs for the co-factors are shown in the following pattern:

$$\begin{vmatrix} + & - & + & - \\ - & + & - & + \\ + & - & + & - \\ - & + & - & + \end{vmatrix}$$

and so

$$\begin{vmatrix} a & e & i & m \\ b & f & j & n \\ c & g & k & p \\ d & h & l & q \end{vmatrix}$$

$$= a\begin{vmatrix} f & j & n \\ g & k & p \\ h & l & q \end{vmatrix} - b\begin{vmatrix} e & i & m \\ g & k & p \\ h & l & q \end{vmatrix} + c\begin{vmatrix} e & i & m \\ f & j & n \\ h & l & q \end{vmatrix} - d\begin{vmatrix} e & i & m \\ f & j & n \\ g & k & p \end{vmatrix}$$

Evidently the labour of calculating a determinant increases very rapidly with increase in order: in the above example of a fourth-order determinant, it is first resolved into four third-order determinants and then each of these has to be evaluated as previously.

THE RULE OF SARRUS

For a third-order determinant there is a very simple rule for evaluation, which may even allow the result to be obtained mentally in some cases. For the determinant

$$\begin{vmatrix} a & d & g \\ b & e & h \\ c & f & i \end{vmatrix}$$

write it out without the vertical lines, repeating the first two columns and adding arrows as shown:

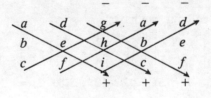

6

We now form products of trios of terms given by the arrows and attach positive signs to products whose arrows point downwards and negative signs where the arrows point upwards:

$$aei + cdh + bfg - ceg - afh - bdi$$

If this expression is compared with the result shown on page 4 it will be seen to be identical. Unfortunately this simple rule only works for third-order determinants.

Matrix inversion

The adjoint matrix

If $\mathbf{A}$ is a square matrix of order greater than 2, then the adjoint of $\mathbf{A}$ (adj. $\mathbf{A}$) is the transpose of the matrix of the co-factors of $\mathbf{A}$. The formation of an adjoint matrix is best demonstrated by means of an example.

Example 1.3   Find the adjoint of the matrix

$$\begin{bmatrix} -1 & 2 & -3 \\ 2 & -1 & 4 \\ 3 & 4 & 1 \end{bmatrix}$$

The co-factor of -1 is $\begin{vmatrix} -1 & 4 \\ 4 & 1 \end{vmatrix} = -1 - 16 = -17$

The co-factor of 2 is $-\begin{vmatrix} 2 & -3 \\ 4 & 1 \end{vmatrix} = -(2 + 12) = -14$

The co-factor of 3 is $\begin{vmatrix} 2 & -3 \\ -1 & 4 \end{vmatrix} = 8 - 3 = 5$

The co-factor of 2 is $-\begin{vmatrix} 2 & 4 \\ 3 & 1 \end{vmatrix} = -(2 - 12) = 10$

The co-factor of -1 is $\begin{vmatrix} -1 & -3 \\ 3 & 1 \end{vmatrix} = -1 + 9 = 8$

The co-factor of 4 is $-\begin{vmatrix} -1 & -3 \\ 2 & 4 \end{vmatrix} = -(-4 + 6) = -2$

The co-factor of -3 is $\begin{vmatrix} 2 & -1 \\ 3 & 4 \end{vmatrix} = 8 + 3 = 11$

The co-factor of 4 is $\quad - \begin{vmatrix} -1 & 2 \\ 3 & 4 \end{vmatrix} \quad = -(-4-6) = 10$

The co-factor of 1 is $\quad \begin{vmatrix} -1 & 2 \\ 2 & -1 \end{vmatrix} \quad = 1 - 4 = -3$

Thus the matrix of co-factors is

$$\begin{bmatrix} -17 & 10 & 11 \\ -14 & 8 & 10 \\ 5 & -2 & -3 \end{bmatrix}$$

The transpose of this is the adjoint of the original matrix

$$\begin{bmatrix} -17 & -14 & 5 \\ 10 & 8 & -2 \\ 11 & 10 & -3 \end{bmatrix}$$

Only square matrices have an adjoint, and the definition given at the beginning of this section applies to third- and higher-order matrices. In the case of a second-order matrix, the adjoint matrix is given by reversing the elements of the leading diagonal and changing the sign of the remaining two elements.

The inverse matrix

We have already seen how to invert a second-order matrix: the adjoint matrix is divided by the determinant. The same procedure is adopted for a matrix of any order.

Example 1.4   Find the inverse matrix of

$$\begin{bmatrix} -1 & 2 & -3 \\ 2 & -1 & 4 \\ 3 & 4 & 1 \end{bmatrix}$$

The adjoint of this matrix has already been found (Example 1.3):

$$\begin{bmatrix} -17 & -14 & 5 \\ 10 & 8 & -2 \\ 11 & 10 & -3 \end{bmatrix}$$

so also has the determinant (Example 1.2), which is 4. Hence the inverse of the matrix is

$$\frac{1}{4}\begin{bmatrix} -17 & -14 & 5 \\ 10 & 8 & -2 \\ 11 & 10 & -3 \end{bmatrix} = \begin{bmatrix} -4\frac{1}{4} & -3\frac{1}{2} & 1\frac{1}{4} \\ 2\frac{1}{2} & 2 & -\frac{1}{2} \\ 2\frac{3}{4} & 2\frac{1}{2} & -\frac{3}{4} \end{bmatrix}$$

The singular matrix

Consider the matrix

$$\begin{bmatrix} 4 & 3 \\ -4 & -3 \end{bmatrix}$$

The adjoint of this matrix is

$$\begin{bmatrix} -3 & -3 \\ 4 & 4 \end{bmatrix}$$

but the determinant is $(4)(-3) - (-4)(3) = -12 - (-12) = 0$. Since an inverse matrix is formed by dividing every element in the adjoint matrix by the determinant, then if the determinant is zero the inverse matrix cannot be formed. A matrix whose determinant is zero is called a singular matrix, and it has no inverse even though it is square.

Example 1.5   Show that

$$\begin{bmatrix} 2 & 3 & -2 \\ 1 & -2 & 3 \\ 4 & -1 & 4 \end{bmatrix}$$

is a singular matrix.

To show that the matrix is singular, we need merely to calculate the determinant and see that it is zero. So

$$\begin{vmatrix} 2 & 3 & -2 \\ 1 & -2 & 3 \\ 4 & -1 & 4 \end{vmatrix} = 2\begin{vmatrix} -2 & 3 \\ -1 & 4 \end{vmatrix} - 1\begin{vmatrix} 3 & -2 \\ -1 & 4 \end{vmatrix} + 4\begin{vmatrix} 3 & -2 \\ -2 & 3 \end{vmatrix}$$

$$= 2(-8+3) - 1(12-2) + 4(9-4)$$

$$= -10 - 10 + 20 = 0$$

Linear equations

Introduction to the solution of linear equations

SOLUTION BY ELIMINATION

Suppose we had to solve the following pair of simultaneous linear equations (linear because both x and y are of the first degree and are in separate terms).

$$3x - 2y = 5 \qquad (1)$$

$$8x - 3y = 30 \qquad (2)$$

The method you know already is an elimination process. Multiply *(1)* by 3 and multiply *(2)* by 2, giving

$$9x - 6y = 15$$

$$16x - 6y = 60$$

Subtraction gives $7x = 45$, and so $x = 6\frac{3}{7}$. Substitute this value of x into one of the two original equations (e.g. the first): $(3)(45/7) - 2y = 5$ which gives $y = 7\frac{1}{7}$.

Such equations can be solved by matrix methods, or even by using determinants alone.

MATRIX SOLUTION

Using the above example, the pair of equations is

$$3x - 2y = 5$$

$$8x - 3y = 30$$

Now this pair of equations can be written in matrix form:

$$\begin{bmatrix} 3 & -2 \\ 8 & -3 \end{bmatrix} \begin{bmatrix} x \\ y \end{bmatrix} = \begin{bmatrix} 5 \\ 30 \end{bmatrix} \qquad (1.2)$$

as can be seen if the matrix multiplication on the left-hand side is carried out. Now, call the matrix of coefficients **A**, the column vector of the two unknowns **x**, and the column vector on the right-hand side **b**; then we have

$$\mathbf{Ax} = \mathbf{b} \qquad (1.3)$$

where (1.3) is a more concise expression of (1.2). Next, pre-multiply both sides of (1.3) by the inverse of **A**:

$$\mathbf{A}^{-1}\mathbf{Ax} = \mathbf{A}^{-1}\mathbf{b}$$

i.e.

$$\mathbf{Ix} = \mathbf{A}^{-1}\mathbf{b}$$

since the product of a matrix and its inverse is the unit matrix. Further, as the product of any matrix and the unit matrix is the matrix itself (i.e. $\mathbf{Ix} = \mathbf{x}$), we have

$$\mathbf{x} = \mathbf{A}^{-1}\mathbf{b} \qquad (1.4)$$

The left-hand side of (1.4) is the column vector of the two unknowns. The right-hand side consists of the product of a known column vector, $\mathbf{b}$, and the inverse of the matrix of known coefficients, $\mathbf{A}$, and so this side is known completely. Hence, the equation can be solved. Applying this to the above numerical example, we first require the inverse of

$$\begin{bmatrix} 3 & -2 \\ 8 & -3 \end{bmatrix}$$

The adjoint is

$$\begin{bmatrix} -3 & 2 \\ -8 & 3 \end{bmatrix}$$

and the determinant is $(-3)(3) - (-8)(2) = 7$; so the inverse is

$$\tfrac{1}{7}\begin{bmatrix} -3 & 2 \\ -8 & 3 \end{bmatrix}$$

Hence

$$\begin{bmatrix} x \\ y \end{bmatrix} = \tfrac{1}{7}\begin{bmatrix} -3 & 2 \\ -8 & 3 \end{bmatrix}\begin{bmatrix} 5 \\ 30 \end{bmatrix}$$

$$= \tfrac{1}{7}\begin{bmatrix} (-3)(5)+(2)(30) \\ (-8)(5)+(3)(30) \end{bmatrix} = \tfrac{1}{7}\begin{bmatrix} 45 \\ 50 \end{bmatrix} = \begin{bmatrix} 6\tfrac{3}{7} \\ 7\tfrac{1}{7} \end{bmatrix}$$

An advantage of the matrix method over the elimination procedure is that the same general formulation can be used for a group of simultaneous equations of any size, so long as the number of unknowns is equal to the number of equations in the set. However large the set, relationships (1.3) and (1.4) always apply.

Example 1.6 Solve, by the matrix method

$$x + 2y + z = 0$$
$$3x + 2y + z = 2$$
$$2x + 3y + 2z = 2$$

11

Put

$$\mathbf{A} = \begin{bmatrix} 1 & 2 & 1 \\ 3 & 2 & 1 \\ 2 & 3 & 2 \end{bmatrix} \quad \mathbf{x} = \begin{bmatrix} x \\ y \\ z \end{bmatrix} \quad \mathbf{b} = \begin{bmatrix} 0 \\ 2 \\ 2 \end{bmatrix}$$

The main problem is finding the inverse of **A**. When this has been achieved, pre-multiplication of **b** by $\mathbf{A}^{-1}$ gives the solution. Now

$$\text{adj } \mathbf{A} = \begin{bmatrix} 1 & -1 & 0 \\ -4 & 0 & 2 \\ 5 & 1 & -4 \end{bmatrix} \quad \text{and} \quad |A| = -2$$

Hence

$$\mathbf{A}^{-1} = -\tfrac{1}{2} \begin{bmatrix} 1 & -1 & 0 \\ -4 & 0 & 2 \\ 5 & 1 & -4 \end{bmatrix}$$

and so

$$\begin{bmatrix} x \\ y \\ z \end{bmatrix} = -\tfrac{1}{2} \begin{bmatrix} 1 & -1 & 0 \\ -4 & 0 & 2 \\ 5 & 1 & -4 \end{bmatrix} \begin{bmatrix} 0 \\ 2 \\ 2 \end{bmatrix} = -\tfrac{1}{2} \begin{bmatrix} -2 \\ 4 \\ -6 \end{bmatrix} = \begin{bmatrix} 1 \\ -2 \\ 3 \end{bmatrix}$$

giving the solutions $x = 1$, $y = -2$, $z = 3$.

DETERMINANT SOLUTION

The method is best illustrated for a pair of equations, and we shall use the previous example:

$$3x - 2y = 5$$
$$8x - 3y = 30$$

Define the following three determinants:

$$\Delta = \begin{vmatrix} 3 & -2 \\ 8 & -3 \end{vmatrix} \quad \Delta_1 = \begin{vmatrix} 5 & -2 \\ 30 & -3 \end{vmatrix} \quad \Delta_2 = \begin{vmatrix} 3 & 5 \\ 8 & 30 \end{vmatrix}$$

The first determinant is formed by the coefficients on the left-hand side of the equation. The second determinant has the coefficients of x in Δ replaced by the column of numbers from the right-hand side of the equations, whereas the third determinant has the coefficients of y in Δ replaced by this column.

12

A result, known as Cramer's rule, states that

$$x = \Delta_1/\Delta \quad \text{and} \quad y = \Delta_2/\Delta$$

Now

$$\Delta = -9 + 16 = 7 \quad \Delta_1 = -15 + 60 = 45 \quad \Delta_2 = 90 - 40 = 50$$

Thus

$$x = \frac{45}{7} = 6\tfrac{3}{7} \quad \text{and} \quad y = \frac{50}{7} = 7\tfrac{1}{7} \text{ (as before)}$$

Cramer's rule is applied in a similar fashion to a larger system of equations.

Example 1.7 Solve the system of equations in Example 1.6, using Cramer's rule.

With three equations in three unknowns, we shall have third-order determinants. Four of these will be required, the first being the determinant of the coefficients on the left-hand side,

$$\Delta = \begin{vmatrix} 1 & 2 & 1 \\ 3 & 2 & 1 \\ 2 & 3 & 2 \end{vmatrix} = -2$$

A second determinant is formed by substituting the values of the right-hand sides of the three equations for the first column of Δ:

$$\Delta_1 = \begin{vmatrix} 0 & 2 & 1 \\ 2 & 2 & 1 \\ 2 & 3 & 2 \end{vmatrix} = -2$$

The third determinant is formed by substituting in the second column of Δ:

$$\Delta_2 = \begin{vmatrix} 1 & 0 & 1 \\ 3 & 2 & 1 \\ 2 & 2 & 2 \end{vmatrix} = 4$$

Finally, a determinant is formed from Δ by substitution in the last column:

$$\Delta_3 = \begin{vmatrix} 1 & 2 & 0 \\ 3 & 2 & 2 \\ 2 & 3 & 2 \end{vmatrix} = -6$$

Then

$$x = \Delta_1/\Delta = -2/-2 = 1$$
$$y = \Delta_2/\Delta = 4/-2 = -2$$
$$z = \Delta_3/\Delta = -6/-2 = 3$$

(as before).

Types of solution

In attempting to solve linear equations, three types of solution are possible. We shall use the elimination method on pairs of equations in two unknowns to illustrate the solution types, in the example below. It is illuminating to give a geometrical interpretation of the results, and a single linear equation in two unknowns represents a straight line in two-dimensional space. For example, the first of the pair of equations on page 10 is

$$3x - 2y = 5$$

which can be rearranged into the more familiar form of the equation of a straight line as

$$y = -2.5 + 1.5x$$

that is, a straight line of gradient 1.5 and intercept -2.5.

Example 1.8   Solve the following systems of equations:

(a) $3x + 2y = 7$ *(b)* $3x + 2y = 7$ *(c)* $3x + 2y = 7$

$2x + 5y = 12$ $6x + 4y = 14$ $6x + 4y = 8$

(a) Multiply the first equation by 2 and the second by 3:

$$6x + 4y = 14$$
$$6x + 15y = 36$$

Subtract the first from the second, which gives $11y = 22$ or $y = 2$. On substituting for y in the first equation, we have $3x + 4 = 7$ or $x = 1$. So the solution is $x = 1$, $y = 2$. The equations can be rewritten as $y = 3.5 - 1.5x$ and $y = 2.4 - 0.4x$. The lines are shown in Figure 1.1a, and the solution is the point of intersection of the lines P(1, 2).

(b) Multiply the first equation by 2 and leave the second one unchanged:

$$6x + 4y = 14$$
$$6x + 4y = 14$$

14

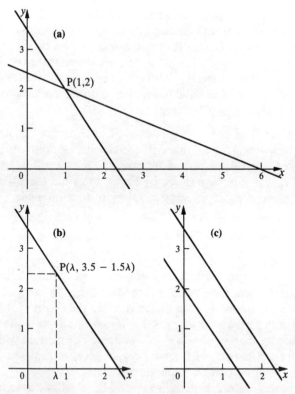

Figure 1.1 (a) Graph of equations $y = 3.5 - 1.5x$ and $y = 2.4 - 0.4x$. (b) Graph of the equation $y = 3.5 - 1.5x$. (c) Graph of the equations $y = 3.5 - 1.5x$ and $y = 2 - 1.5x$. (See Example 1.8.)

Subtraction gives $0 = 0$. This result is perfectly sound, but it shows that one of the two equations is redundant. Indeed, inspection shows that the second equation is exactly twice the first one, so in effect the two equations represent a single straight line $y = 3.5 - 1.5x$. However, another way of viewing the situation is to regard the systems as comprising two coincident lines. This means that there are an infinity of solutions, but for any particular value of x there is, however, a unique value of y. For instance, let $x = \lambda$; then, substituting into $3x + 2y = 7$ or $y = 3.5 - 1.5x$ gives $y = 3.5 - 1.5\lambda$. So a typical solution, shown in Figure 1.1b, is the point $P(\lambda, 3.5 - 1.5\lambda)$.

(c) Again, multiply the first equation by 2 and leave the second unchanged:

$$6x + 4y = 14$$

$$6x + 4y = 8$$

15

Subtraction now gives the anomalous result $0 = 6$. We have to conclude that there is no solution to this pair of equations; the pair are said to be **inconsistent**. Rearrangement of the two equations gives $y = 3.5 - 1.5x$ and $y = 2 - 1.5x$. The gradients of the two lines are the same, but their intercepts differ. Consequently the two equations represent a pair of parallel lines, and so no intersection or solution of the equations is possible (Fig. 1.1c).

The three kinds of solution encountered in Example 1.8, namely, (1) a single unique solution, (2) an infinity of solutions, (3) no solution, may arise in a system of linear equations of any size. In the case of just two equations in two unknowns it is easy to see by inspection what the situation is, but for any larger system an attempt at solution must be made.

Solution by the Gauss–Jordan elimination method

Despite the possibility of solving linear equations by matrices and determinants, the use of an elimination method is usually much simpler in practice, especially where the equations are solved using a computer program. The commonest elimination algorithm is called the Gauss–Jordan method. To illustrate the process with two equations in two unknowns is rather trivial; hence we shall employ a 3 by 3 system. The method will be laid out algebraically first so that, with some perseverance, you can see exactly what is going on; then numerical examples will be given.

At this stage, before proceeding with the algebraic statement of the Gauss–Jordan method, we must introduce the standard general notation for linear equations.

Recall that a system of linear equations can be written succinctly in matrix form as

$$\mathbf{Ax} = \mathbf{b} \tag{1.5}$$

For three equations in three unknowns, the details of the matrices in (1.5) are as follows:

$$\mathbf{A} = \begin{bmatrix} a_{11} & a_{12} & a_{13} \\ a_{21} & a_{22} & a_{23} \\ a_{31} & a_{32} & a_{33} \end{bmatrix} \qquad \mathbf{x} = \begin{bmatrix} x_1 \\ x_2 \\ x_3 \end{bmatrix} \qquad \mathbf{b} = \begin{bmatrix} b_1 \\ b_2 \\ b_3 \end{bmatrix}$$

where the elements a_{ij} of matrix $\mathbf{A}$ and the elements b_i of column vector $\mathbf{b}$ are known numerical values, and the elements x_i of column vector $\mathbf{x}$ are the unknowns requiring to be found. Writing out (1.5) in full gives the required equations:

$$a_{11}x_1 + a_{12}x_2 + a_{13}x_3 = b_1 \tag{1}$$
$$a_{21}x_1 + a_{22}x_2 + a_{23}x_3 = b_2 \tag{2}$$
$$a_{31}x_1 + a_{32}x_2 + a_{33}x_3 = b_3 \tag{3}$$

Now for the Gauss–Jordan method. Divide *(1)* by a_{11} (to obtain the unit coefficient for x_1):

$$x_1 + \frac{a_{12}}{a_{11}} x_2 + \frac{a_{13}}{a_{11}} x_3 = \frac{b_1}{a_{11}} \qquad (1a)$$

Multiply *(1a)* by $- a_{21}$:

$$- a_{21}x_1 - \frac{a_{21}a_{12}}{a_{11}} x_2 - \frac{a_{21}a_{13}}{a_{11}} x_3 = - \frac{a_{21}b_1}{a_{11}} \qquad (1b)$$

Add *(1b)* to *(2)*, giving *(2a)*, and write out the modified system:

$$x_1 + \qquad \frac{a_{12}}{a_{11}} x_2 + \qquad \frac{a_{13}}{a_{11}} x_3 = \frac{b_1}{a_{11}} \qquad (1a)$$

$$\left(a_{22} - \frac{a_{21}a_{12}}{a_{11}}\right) x_2 + \left(a_{23} - \frac{a_{21}a_{13}}{a_{11}}\right) x_3 = b_2 - \frac{a_{21}b_1}{a_{11}} \qquad (2a)$$

$$a_{31}x_1 + \qquad a_{32}x_2 + \qquad a_{33}x_3 = b_3 \qquad (3)$$

and we see that x_1 has been eliminated from the second equation. Now multiply *(1a)* by $- a_{31}$:

$$- a_{31}x_1 - \frac{a_{31}a_{12}}{a_{11}} x_2 - \frac{a_{31}a_{13}}{a_{11}} x_3 = - \frac{a_{31}b_1}{a_{11}} \qquad (1c)$$

Add *(1c)* to *(3)*, giving *(3a)*, and write out the new system:

$$x_1 + \qquad \frac{a_{12}}{a_{11}} x_2 + \qquad \frac{a_{13}}{a_{11}} x_3 = \frac{b_1}{a_{11}} \qquad (1a)$$

$$\left(a_{22} - \frac{a_{21}a_{12}}{a_{11}}\right) x_2 + \left(a_{23} - \frac{a_{21}a_{13}}{a_{11}}\right) x_3 = b_2 - \frac{a_{21}b_1}{a_{11}} \qquad (2a)$$

$$\left(a_{32} - \frac{a_{31}a_{12}}{a_{11}}\right) x_2 + \left(a_{33} - \frac{a_{31}a_{13}}{a_{11}}\right) x_3 = b_3 - \frac{a_{31}b_1}{a_{11}} \qquad (3a)$$

We have now eliminated x_1 from all equations except the first, and the aim now is to eliminate x_2 from the third equation by a similar process. To prevent the coefficients from getting too clumsy, rewrite the above system as

$$x_1 + a'_{12}x_2 + a'_{13}x_3 = b'_1 \qquad (1a)$$

$$a'_{22}x_2 + a'_{23}x_3 = b'_2 \qquad (2a)$$

$$a'_{32}x_2 + a'_{33}x_3 = b'_3 \qquad (3a)$$

where

$$a'_{1j} = \frac{a_{1j}}{a_{11}} \qquad j = 2, 3$$

$$a'_{ij} = a_{ij} - \frac{a_{i1}a_{1j}}{a_{11}} \qquad i = 2, 3 \qquad j = 1, \ldots, 3$$

$$b'_1 = \frac{b_1}{a_{11}} \qquad b'_i = b_i - \frac{a_{i1}b_1}{a_{11}} \qquad i = 2, 3$$

Divide *(2a)* by a'_{22}:

$$x_2 + \frac{a'_{23}}{a'_{22}} x_3 = \frac{b'_2}{a'_{22}} \tag{2b}$$

Multiply *(2b)* by $- a'_{32}$:

$$- a'_{32}x_2 - \frac{a'_{32}a'_{23}}{a'_{22}} x_3 = - \frac{a'_{32}b'_2}{a'_{22}} \tag{2c}$$

Add *(2c)* to *(3a)*, giving *(3b)*, and write out the new systems:

$$x_1 + a'_{12}x_2 + \qquad\qquad a'_{13}x_3 = b'_1 \tag{1a}$$

$$x_2 + \qquad \frac{a'_{23}}{a'_{22}} x_3 = \frac{b'_2}{a'_{22}} \tag{2b}$$

$$\left(a'_{33} - \frac{a'_{32}a'_{23}}{a'_{22}}\right) x_3 = b'_3 - \frac{a'_{32}b'_2}{a'_{22}} \tag{3b}$$

Put each side of *(3b)* over a common denominator, a'_{22}, which then cancels, giving

$$(a'_{22}a'_{33} - a'_{32}a'_{23})x_3 = a'_{22}b'_3 - a'_{32}b'_2$$

So the system finally becomes

$$x_1 + a'_{12}x_2 + a'_{13}x_3 = b'_1 \tag{1a}$$

$$x_2 + \frac{a'_{23}}{a'_{22}} x_3 = \frac{b'_2}{a'_{22}} \tag{2b}$$

$$x_3 = \frac{a'_{22}b'_3 - a'_{32}b'_2}{a'_{22}a'_{33} - a'_{32}a'_{23}} \tag{3c}$$

The solution for x_3 is given directly from *(3c)*, then substitution of x_3 in *(2b)* gives the solution for x_2, and finally, inserting the values for x_2 and x_3 in *(1a)* yields the solution for x_1.

It would not be sensible to try to utilise *(1a)*, *(2b)*, and *(3c)* directly by 'plugging in' values for the a_{ij} and b_j via the conversion formulae beneath the system *(1a)*, *(2a)*, and *(3a)*. Instead, for actual sets of equations, the procedure given above is followed through without change, as the following example shows.

Example 1.9 *Solve the following systems of linear equations*

(a)
$$2x_1 + 5x_2 + 2x_3 = 6 \tag{1}$$
$$3x_1 + x_2 + 7x_3 = 0 \tag{2}$$
$$-4x_1 - 3x_2 + 4x_3 = 10 \tag{3}$$

(b)
$$2x_1 + x_2 + 2x_3 - x_4 = 0 \tag{1}$$
$$6x_1 + 8x_2 + 12x_3 - 13x_4 = -21 \tag{2}$$
$$10x_1 + 2x_2 + 2x_3 + 3x_4 = 21 \tag{3}$$
$$-4x_1 + 4x_2 + x_3 - 3x_4 = -9 \tag{4}$$

(a) Divide *(1)* by 2:
$$x_1 + \tfrac{5}{2}x_2 + x_3 = 3 \tag{1a}$$

Multiply *(1a)* by -3:
$$-3x_1 - \tfrac{15}{2}x_2 - 3x_3 = -9 \tag{1b}$$

Add *(1b)* to *(2)*, giving *(2a)*, and write out the new system:
$$x_1 + \tfrac{5}{2}x_2 + x_3 = 3 \tag{1a}$$
$$-\tfrac{13}{2}x_2 + 4x_3 = -9 \tag{2a}$$
$$-4x_1 - 3x_2 + 4x_3 = 10 \tag{3}$$

Multiply *(1a)* by 4:
$$4x_1 + 10x_2 + 4x_3 = 12 \tag{1c}$$

Add *(1c)* to *(3)*, giving *(3a)*, and write out the new system:
$$x_1 + \tfrac{5}{2}x_2 + x_3 = 3 \tag{1a}$$
$$-\tfrac{13}{2}x_2 + 4x_3 = -9 \tag{2a}$$
$$7x_2 + 8x_3 = 22 \tag{3a}$$

Divide *(2a)* by $-\tfrac{13}{2}$
$$x_2 - \tfrac{8}{13}x_3 = \tfrac{18}{13} \tag{2b}$$

Multiply *(2b)* by -7:
$$-7x_2 + \tfrac{56}{13}x_3 = -\tfrac{126}{13} \tag{2c}$$

19

Add *(2c)* to *(3a)*, giving *(3b)*, and write out the new system:

$$x_1 + \tfrac{5}{2} x_2 + \quad x_3 = 3 \qquad (1a)$$

$$x_2 - \tfrac{8}{13} x_3 = \tfrac{18}{13} \qquad (2b)$$

$$\tfrac{160}{13} x_3 = \tfrac{160}{13} \qquad (3b)$$

From *(3b)*, $x_3 = 1$. Substituting into *(2b)* gives $x_2 - \tfrac{8}{13} = \tfrac{18}{13}$, and so $x_2 = 2$. Finally, substitution into *(1a)* gives $x_1 + 5 + 1 = 3$, yielding $x_1 = -3$. A useful way of writing out the solution of linear equations is in the form of a column vector:

$$\begin{bmatrix} x_1 \\ x_2 \\ x_3 \end{bmatrix} = \begin{bmatrix} -3 \\ 2 \\ 1 \end{bmatrix}$$

and so the solution can be expressed in the form

$$\mathbf{x} = \mathbf{p}$$

(b) Divide *(1)* by 2:

$$x_1 + \tfrac{1}{2} x_2 + x_3 - \tfrac{1}{2} x_4 = 0 \qquad (1a)$$

Multiply *(1a)* by -6:

$$-6x_1 - 3x_2 - 6x_3 + 3x_4 = 0 \qquad (1b)$$

Add *(1b)* to *(2)*, giving *(2a)*, and write out the new system:

$$x_1 + \tfrac{1}{2} x_2 + x_3 - \tfrac{1}{2} x_4 = 0 \qquad (1a)$$

$$5x_2 + 6x_3 - 10x_4 = -21 \qquad (2a)$$

$$10x_1 + 2x_2 + 2x_3 + 3x_4 = 21 \qquad (3)$$

$$-4x_1 + 4x_2 + x_3 - 3x_4 = -9 \qquad (4)$$

Multiply *(1a)* by -10:

$$-10x_1 - 5x_2 - 10x_3 + 5x_4 = 0 \qquad (1c)$$

Add *(1c)* to *(3)*, giving *(3a)*, and write out the new system:

$$x_1 + \tfrac{1}{2} x_2 + x_3 - \tfrac{1}{2} x_4 = 0 \qquad (1a)$$

$$5x_2 + 6x_3 - 10x_4 = -21 \qquad (2a)$$

$$-3x_2 - 8x_3 + 8x_4 = 21 \qquad (3a)$$

$$-4x_1 + 4x_2 + x_3 - 3x_4 = -9 \qquad (4)$$

Multiply *(1a)* by 4:

$$4x_1 + 2x_2 + 4x_3 - 2x_4 = 0 \qquad (1d)$$

Add *(1d)* to *(4)*, giving *(4a)*, and write out the new system:

$$x_1 + \tfrac{1}{2}x_2 + x_3 - \tfrac{1}{2}x_4 = 0 \tag{1a}$$

$$5x_2 + 6x_3 - 10x_4 = -21 \tag{2a}$$

$$-3x_2 - 8x_3 + 8x_4 = 21 \tag{3a}$$

$$6x_2 + 5x_3 - 5x_4 = -9 \tag{4a}$$

Divide *(2a)* by 5:

$$x_2 + \tfrac{6}{5}x_3 - 2x_4 = -\tfrac{21}{5} \tag{2b}$$

Multiply *(2b)* by 3:

$$3x_2 + \tfrac{18}{5}x_3 - 6x_4 = -\tfrac{63}{5} \tag{2c}$$

Add *(2c)* to *(3a)*, giving *(3b)*, and write out the new system:

$$x_1 + \tfrac{1}{2}x_2 + \quad x_3 - \tfrac{1}{2}x_4 = 0 \tag{1a}$$

$$x_2 + \quad\tfrac{6}{5}x_3 - 2x_4 = -\tfrac{21}{5} \tag{2b}$$

$$-\tfrac{22}{5}x_3 + 2x_4 = \tfrac{42}{5} \tag{3b}$$

$$6x_2 + \quad 5x_3 - 5x_4 = -9 \tag{4a}$$

Multiply *(2b)* by -6:

$$-6x_2 - \tfrac{36}{5}x_3 + 12x_4 = \tfrac{126}{5} \tag{2d}$$

Add *(2d)* to *(4a)*, giving *(4b)*, and write out the new system:

$$x_1 + \tfrac{1}{2}x_2 + \quad x_3 - \tfrac{1}{2}x_4 = 0 \tag{1a}$$

$$x_2 + \quad\tfrac{6}{5}x_3 - 2x_4 = -\tfrac{21}{5} \tag{2b}$$

$$-\tfrac{22}{5}x_3 + 2x_4 = \tfrac{42}{5} \tag{3b}$$

$$-\tfrac{11}{5}x_3 + 7x_4 = \tfrac{81}{5} \tag{4b}$$

Divide *(3b)* by $-\tfrac{22}{5}$:

$$x_3 - \tfrac{5}{11}x_4 = -\tfrac{21}{11} \tag{3c}$$

Multiply *(3c)* by $\tfrac{11}{5}$:

$$\tfrac{11}{5}x_3 - x_4 = -\tfrac{21}{5} \tag{3d}$$

Add *(3d)* to *(4b)*, giving *(4c)*, and write out the new system:

$$x_1 + \tfrac{1}{2}x_2 + \quad x_3 - \quad \tfrac{1}{2}x_4 = 0 \tag{1a}$$

$$x_2 + \tfrac{6}{5}x_3 - \quad 2x_4 = -\tfrac{21}{5} \tag{2b}$$

$$x_3 - \tfrac{5}{11}x_4 = -\tfrac{21}{11} \tag{3c}$$

$$6x_4 = 12 \tag{4c}$$

21

From *(4c)*, $x_4 = 2$. Substituting into *(3c)* gives $x_3 - \frac{10}{11} = -\frac{21}{11}$, and so $x_3 = -1$; substituting into *(2b)* gives $x_2 - \frac{6}{5} - 4 = -\frac{21}{5}$, yielding $x_2 = 1$. Finally, substitution into *(1a)* gives $x_1 + \frac{1}{2} - 1 - 1 = 0$, and so $x_1 = 1\frac{1}{2}$. In vector form:

$$\begin{bmatrix} x_1 \\ x_2 \\ x_3 \\ x_4 \end{bmatrix} = \begin{bmatrix} 1\frac{1}{2} \\ 1 \\ -1 \\ 2 \end{bmatrix}$$

THE 'ELEMENTARY TRANSFORMATIONS'

In the above examples the only operations performed have been combinations of the following.

(a) Multiplying an equation by a non-zero number.
(b) Adding a multiple of one equation to another equation of the system.

The division of equation *(1)* of a system by a_{11} (the coefficient of x_1) is the same as multiplying by $1/a_{11}$, which satisfies (a) above. The occasional subtraction which appeared in the above example is equivalent to addition; it merely saves a prior operation of multiplying by -1 before doing the addition.

The two operations are known as the **elementary transformations**, and any system of linear equations can be solved by using them alone in any desired or necessary combination. The important property of these transformations is that however many times they are used, and in whatever combination, they do not change the *underlying* relationships within the *original* system.

Non-degenerative systems

In the systems considered in the previous section the number of unknowns equals the number of equations (say n of each), and the matrices of the coefficients on the left-hand sides are non-singular (see Exercise 1.2 at end of chapter). Such a system is known as a **non-degenerative system of order** n and always has one and only one solution. Another convenient way of classifying linear equations is according to whether the column vector on the right-hand side is zero or not. If the system can be represented in matrix form as

$$\mathbf{Ax} = \mathbf{0}$$

the system is said to be **homogeneous**, but if the matrix representation is

$$\mathbf{Ax} = \mathbf{b} \qquad \mathbf{b} \neq \mathbf{0}$$

22

the system is **inhomogeneous**. A non-degenerative homogeneous system only has the trivial solution $\mathbf{x} = \mathbf{0}$, i.e. all the $x_i = 0$.

Non-degenerative inhomogeneous systems are undoubtedly the most important kind of linear equations arising in practice, but the other kinds must now be examined in order for you to know something of the problems that can occur, and their mode of solution if such systems are encountered.

General systems

THE DEGENERATE HOMOGENEOUS SYSTEM

In the matrix representation of an homogeneous system of linear equations, $\mathbf{Ax} = \mathbf{0}$, the word **degenerate** implies that $\mathbf{A}$ is singular.

Example 1.10   Solve the system

$$2x_1 + 3x_2 - 2x_3 = 0 \tag{1}$$

$$x_1 - 2x_2 + 3x_3 = 0 \tag{2}$$

$$4x_1 - x_2 + 4x_3 = 0 \tag{3}$$

The matrix of coefficients is identical to that of Example 1.5, which was shown to be singular.

First, interchange rows *(1)* and *(2)* to avoid introducing fractions into *(1)* on division by 2 (the interchanging of rows may also be regarded as an elementary transformation, since nothing is changed in the system as a whole).

$$x_1 - 2x_2 + 3x_3 = 0 \tag{2}$$

$$2x_1 + 3x_2 - 2x_3 = 0 \tag{1}$$

$$4x_1 - x_2 + 4x_3 = 0 \tag{3}$$

Multiply *(2)* by -2:

$$-2x_1 + 4x_2 - 6x_3 = 0 \tag{2a}$$

Add *(2a)* to *(1)*, giving *(1a)*, and write out the new system:

$$x_1 - 2x_2 + 3x_3 = 0 \tag{2}$$

$$7x_2 - 8x_3 = 0 \tag{1a}$$

$$4x_1 - x_2 + 4x_3 = 0 \tag{3}$$

Multiply *(2)* by -4:

$$-4x_1 + 8x_2 - 12x_3 = 0 \tag{2b}$$

Add *(2b)* to *(3)*, giving *(3a)*, and write out the new system:

$$x_1 - 2x_2 + 3x_3 = 0 \tag{2}$$

$$7x_2 - 8x_3 = 0 \tag{1a}$$

$$7x_2 - 8x_3 = 0 \tag{3a}$$

Inspection shows that *(3a)* is redundant because it is identical to *(1a)*. A more formal way of showing this would be: subtract *(1a)* from *(3a)*, obtaining

$$0 = 0 \tag{3b}$$

a valid, but uninformative relationship (see page 15). So we are left with *(2)* and *(1a)*.

From *(1a)* we have $x_2 = \frac{8}{7} x_3$. Substituting into *(2)* gives $x_1 - \frac{16}{7} x_3 + 3x_3 = 0$, and so $x_1 = -\frac{5}{7} x_3$. Clearly, having one fewer equations than unknowns provides insufficient information for an explicit solution. The value of one unknown has to be fixed before the others can be evaluated. Above, we have expressed x_1 and x_2 in terms of x_3; so let $x_3 = \lambda$, then $x_2 = \frac{8}{7} \lambda$ and $x_1 = -\frac{5}{7} \lambda$. In column-vector form, the complete solution is

$$\begin{bmatrix} x_1 \\ x_2 \\ x_3 \end{bmatrix} = \lambda \begin{bmatrix} -\frac{5}{7} \\ \frac{8}{7} \\ 1 \end{bmatrix}$$

Evidently the solution can be expressed in the form

$$\mathbf{x} = \lambda \mathbf{p}$$

where λ can be *any* multiplier, and p gives the ratio $x_1 : x_2 : x_3$ that must be present in every solution. For example,

$$\begin{bmatrix} -5 \\ 8 \\ 7 \end{bmatrix} \text{ is a solution, } \quad \text{so is } \begin{bmatrix} 10 \\ -16 \\ -14 \end{bmatrix}, \quad \text{so also is } \begin{bmatrix} 1 \\ -\frac{8}{5} \\ -\frac{7}{5} \end{bmatrix}$$

THE DEGENERATE INHOMOGENEOUS SYSTEM

Again, in terms of $\mathbf{Ax} = \mathbf{b}$, $\mathbf{A}$ is singular, but $\mathbf{b} \neq 0$.

Example 1.11   Solve the system

$$2x_1 + 3x_2 - 2x_3 = 2 \tag{1}$$

$$x_1 - 2x_2 + 3x_3 = 6 \tag{2}$$

$$4x_1 - x_2 + 4x_3 = 14 \tag{3}$$

The left-hand side of this system is identical with that of the system in Example 1.10; only the column vector of the right-hand side differs. This means that the Gauss–Jordan method proceeds in an identical way down as far as

$$x_1 - 2x_2 + 3x_3 = 6 \tag{2}$$

$$7x_2 - 8x_3 = -10 \tag{1a}$$

$$0 = 0 \tag{3b}$$

Continuing as before, *(1a)* gives $x_2 = \frac{8}{7}x_3 - \frac{10}{7}$. Substituting into *(2)* yields

$$x_1 - 2(\tfrac{8}{7}x_3 - \tfrac{10}{7}) + 3x_3 = 6$$

and so $x_1 = -\frac{5}{7}x_3 + \frac{22}{7}$. Putting $x_3 = \lambda$ (a particular value) gives the solution

$$x_1 = -\tfrac{5}{7}\lambda + \tfrac{22}{7}$$

$$x_2 = \tfrac{8}{7}\lambda - \tfrac{10}{7}$$

$$x_3 = \lambda$$

In vector form, the solution is

$$\begin{bmatrix} x_1 \\ x_2 \\ x_3 \end{bmatrix} = \lambda \begin{bmatrix} -\frac{5}{7} \\ \frac{8}{7} \\ 1 \end{bmatrix} + \begin{bmatrix} \frac{22}{7} \\ -\frac{10}{7} \\ 0 \end{bmatrix}$$

So the solution can be expressed in the form

$$\mathbf{x} = \lambda\mathbf{p} + \mathbf{q}$$

Clearly the solution of the system in Example 1.11 is the same as that in Example 1.10 with the addition of a vector of constants, $\mathbf{q}$; this is an important general rule.

If we have a solution of a homogeneous system of equations, then the solution of a corresponding (identical left-hand side) inhomogeneous system (provided that it is consistent) is the same as that of the homogeneous system plus a vector of constants.

That a degenerate inhomogeneous system of equations can easily be inconsistent is shown by the following example, which has the same left-hand side as the previous example, but a slightly different right-hand vector.

Example 1.12   Investigate the system

$$2x_1 + 3x_2 - 2x_3 = 1 \qquad\qquad (1)$$

$$x_1 - 2x_2 + 3x_3 = 6 \qquad\qquad (2)$$

$$4x_1 - x_2 + 4x_3 = 14 \qquad\qquad (3)$$

The procedure may be followed as in the previous two examples down to the following set:

$$x_1 - 2x_2 + 3x_3 = 6 \qquad\qquad (2)$$

$$7x_2 - 8x_3 = -11 \qquad\qquad (1a)$$

$$7x_2 - 8x_3 = 10 \qquad\qquad (3a)$$

Clearly, by inspection, *(1a)* and *(3a)* cannot both be correct, and this inconsistency can be written down formally by subtracting *(1a)* from *(3a)* and writing down the final system.

$$x_1 - 2x_2 + 3x_3 = 6 \qquad\qquad (2)$$

$$7x_2 - 8x_3 = -11 \qquad\qquad (1a)$$

$$0 = 21 \qquad\qquad (3b)$$

Finally, in a consistent degenerate system with more than three unknowns, the solution may contain more than one parameter, as the following example shows.

Example 1.13   Solve the system

$$3x_1 + x_2 + 2x_3 + 4x_4 = 16 \qquad\qquad (1)$$

$$5x_1 + 2x_2 + 3x_3 + 6x_4 = 27 \qquad\qquad (2)$$

$$4x_1 + x_2 + 3x_3 + 6x_4 = 21 \qquad\qquad (3)$$

$$5x_1 + x_2 + 4x_3 + 8x_4 = 26 \qquad\qquad (4)$$

Divide *(1)* by 3:

$$x_1 + \tfrac{1}{3}x_2 + \tfrac{2}{3}x_3 + \tfrac{4}{3}x_4 = \tfrac{16}{3} \qquad\qquad (1a)$$

Multiply *(1a)* by -5, then by -4:

$$-5x_1 - \tfrac{5}{3}x_2 - \tfrac{10}{3}x_3 - \tfrac{20}{3}x_4 = -\tfrac{80}{3} \qquad\qquad (1b)$$

$$-4x_1 - \tfrac{4}{3}x_2 - \tfrac{8}{3}x_3 - \tfrac{16}{3}x_4 = -\tfrac{64}{3} \qquad\qquad (1c)$$

Add *(1b)* to *(2)* and *(4)*, giving *(2a)* and *(4a)*; add *(1c)* to *(3,)* giving *(3a)*;

and write out the new system:

$$x_1 + \tfrac{1}{3}x_2 + \tfrac{2}{3}x_3 + \tfrac{4}{3}x_4 = \tfrac{16}{3} \qquad (1a)$$

$$\tfrac{1}{3}x_2 - \tfrac{1}{3}x_3 - \tfrac{2}{3}x_4 = \tfrac{1}{3} \qquad (2a)$$

$$-\tfrac{1}{3}x_2 + \tfrac{1}{3}x_3 + \tfrac{2}{3}x_4 = -\tfrac{1}{3} \qquad (3a)$$

$$-\tfrac{2}{3}x_2 + \tfrac{2}{3}x_3 + \tfrac{4}{3}x_4 = -\tfrac{2}{3} \qquad (4a)$$

Add *(2a)* to *(3a)*, giving *(3b)*, eliminate fractions from *(2a)* and *(4a)*, rearrange, and write out the revised system:

$$x_1 + \tfrac{1}{3}x_2 + \tfrac{2}{3}x_3 + \tfrac{4}{3}x_4 = \tfrac{16}{3} \qquad (1a)$$

$$x_2 - x_3 - 2x_4 = 1 \qquad (2a)$$

$$-2x_2 + 2x_3 + 4x_4 = -2 \qquad (4a)$$

$$0 = 0 \qquad (3b)$$

Multiply *(2a)* by 2:

$$2x_2 - 2x_3 - 4x_4 = 2 \qquad (2b)$$

Add *(2b)* to *(4a)* and write out the new system:

$$x_1 + \tfrac{1}{3}x_2 + \tfrac{2}{3}x_3 + \tfrac{4}{3}x_4 = \tfrac{16}{3} \qquad (1a)$$

$$x_2 - x_3 - 2x_4 = 1 \qquad (2a)$$

From *(2a)*

$$x_2 = x_3 + 2x_4 + 1$$

Substituting for x_2 in *(1a)* gives

$$x_1 + \tfrac{1}{3}(x_3 + 2x_4 + 1) + \tfrac{2}{3}x_3 + \tfrac{4}{3}x_4 = \tfrac{16}{3}$$

from which $x_1 = -x_3 - 2x_4 + 5$. Putting $x_3 = \lambda$ and $x_4 = \mu$, we have the solution

$$x_1 = -\lambda - 2\mu + 5$$

$$x_2 = \lambda + 2\mu + 1$$

$$x_3 = \lambda$$

$$x_4 = \mu$$

In vector form

$$\begin{bmatrix} x_1 \\ x_2 \\ x_3 \\ x_4 \end{bmatrix} = \lambda \begin{bmatrix} -1 \\ 1 \\ 1 \\ 0 \end{bmatrix} + \mu \begin{bmatrix} -2 \\ 2 \\ 0 \\ 1 \end{bmatrix} + \begin{bmatrix} 5 \\ 1 \\ 0 \\ 0 \end{bmatrix}$$

or

$$\mathbf{x} = \lambda\mathbf{p} + \mu\mathbf{q} + \mathbf{r}$$

As is seen from the expressions for x_1, x_2, x_3, and x_4, we are expressing two of the unknowns (x_1, x_2) in terms of the other two (x_3, x_4), together with scalar constants (5 and 1). If the system had been homogeneous, there would have been no scalar constants, i.e. $\mathbf{r} = \mathbf{0}$.

MORE UNKNOWNS THAN EQUATIONS

This case is always a degenerate system and can never be inconsistent. We have met this case already in the solution of degenerate systems, in Examples 1.10 and 1.11 (eqns *(1a)* and *(2)*), and 1.13 (eqns *(1a)* and *(2a)*). So the last parts of these examples show how to solve equations of this type. No explicit solution is possible, but m of the n unknowns can be expressed in terms of the remaining unknowns, where m is the number of independent equations.

FEWER UNKNOWNS THAN EQUATIONS

In this case, either the system is inconsistent or one or more of the equations is redundant; if the latter, the number of redundant equations is precisely the excess number of equations over the number of unknowns. Pursuance of the Gauss–Jordan procedure will enable the system to be analysed, and a solution to be obtained if the system is not inconsistent.

Ill-conditioned equations

If, in a system of equations $\mathbf{Ax} = \mathbf{b}$, $|A|$ is not exactly zero but is small compared with the elements of $\mathbf{A}$ there may be difficulty in solving the equation. The following example, from Nicolson *et al.* (1961) is illuminating. Consider the equation set

$$1.53x_1 + 3.24x_2 - 4.18x_3 = 20.55$$

$$3.41x_1 - 1.62x_2 - 0.41x_3 = 1.40$$

$$0.88x_1 + 19.44x_2 - 20.08x_3 = 100.00$$

The solution is $x_1 = 1$, $x_2 = 2$, $x_3 = -3$ exactly, as can be verified by substitution. However, as with most numerical procedures in practice, actually to derive the accurate solution depends on the method used and the accuracy of the electronic aids available. Thus, employing Cramer's method on a 10-digit calculator, the exact solution was obtained, but the Gauss–Jordan method gave the solution set $x_1 = 1.000\ 001\ 6$, $x_2 = 2.000\ 002\ 69$, $x_3 = 2.999\ 997\ 328$.

However, there is a more serious problem if all the numbers in the above equations are not exact. If the equations were the result of experimental data, the numbers would be unique to that experiment; a repeat of the experiment would give slightly different numbers. If the 100 on the right-hand side of the last equation is increased by only 0.05%, to 100.05, the solution of the equation set becomes $x_1 = 2.00$, $x_2 = 3.68$, $x_3 = 1.33$. From the experimental point of view the equations are useless and might as well be incompatible; at best they give the order of magnitude of the unknowns. The reason is that the determinant of the coefficients is only -0.405 – less than any of the individual coefficients and much less than most of them; from this viewpoint the matrix is *almost* singular. Although algebraically a matrix is either singular or non-singular, in numerical applications this distinction is too rigid and there is an intermediate class such as that of the above example, called ill-conditioned.

Part of the skill in the design of experiments is to avoid getting results in the form of ill-defined equations. One must ensure that each equation represents a distinctly new piece of information, not just apparently different from the other equations because of experimental errors.

The practical solution of linear equations

Nowadays most sets of linear equations and matrix inversions are done by computer, usually by a process based on the Gauss–Jordan method. Matrix inversion can also be done by what is an extension of the Gauss–Jordan procedure, in which case it is usually called the row–eschelon method. Computer procedures for these purposes are readily available but, as is always the case when using standard computer software, pitfalls abound. In particular, the remarks of the previous section must be borne in mind, and the determinant of the coefficients of the set of equations should always be obtained and compared with the magnitudes of the coefficients themselves. If the determinant is small, the solution must be treated with caution. Also beware of the situation where coefficients differ by many orders of magnitude; rounding errors can become large here. If a double precision facility exists, whereby the number of digits of the mantissas of real numbers can be increased, use it.

If the system or matrix is small, and no computer facilities are to hand, direct calculation using a calculator may be feasible. The problem with two equations in two unknowns is trivial. For three equations in three unknowns, Cramer's rule combined with the Rule of Sarrus for evaluating the determinants is rapid and quite accurate. The Gauss–Jordan method takes much longer, but must be resorted to if the determinant of coefficients, $|A|$, is found to be zero. The latter method must also be used if the set contains more unknowns than there are equations. For a system of four equations, the largest that should be tackled by hand, the Gauss–Jordan method is better than Cramer's rule in terms of time and effort required.

Eigenvalues and eigenvectors

The topics to be presented in this section, although based on a seemingly trivial idea, are outstandingly important in the theory and application of matrices. Before proceeding to the main topic, however, a short introductory section to set the scene is required.

Linear transformations

The matrix product

$$\mathbf{Ax} = \mathbf{b} \tag{1.6}$$

where $\mathbf{A}$ is a square matrix of order n containing known values, $\mathbf{x}$ is an $(n \times 1)$ column vector of known values, and $\mathbf{b}$ is an $(n \times 1)$ column vector containing the result, is known as a linear transformation (in this case of $\mathbf{x}$). The form of (1.6) resembles a system of linear equations, but is simpler to operate on since the whole of the left-hand side is known and merely involves a matrix multiplication.

The important point at issue now is not simply that (1.6) represents a matrix multiplication, but the geometrical interpretation of this multiplication. Because both $\mathbf{x}$ and $\mathbf{b}$ are column vectors they represent vectors in n-dimensional space from the origin to the point whose co-ordinates are the elements of the column vector. Evidently multiplication by $\mathbf{A}$ changes or transforms vector $\mathbf{x}$ into another vector $\mathbf{b}$. For example, let

$$\mathbf{x} = \begin{bmatrix} 2 \\ 1 \end{bmatrix} \quad \text{and} \quad \mathbf{A} = \begin{bmatrix} 1 & 2 \\ 3 & -1 \end{bmatrix}$$

then

$$\begin{bmatrix} 1 & 2 \\ 3 & -1 \end{bmatrix} \begin{bmatrix} 2 \\ 1 \end{bmatrix} = \begin{bmatrix} 4 \\ 5 \end{bmatrix}$$

The situation is shown in Figure 1.2a, where vector $\overrightarrow{OP}$ is transformed to vector $\overrightarrow{OP}'$. For a (3×3) transformation matrix and (3×1) column vector, we have an analogous situation in three-dimensional space; also, similarly, for matrices, vectors, and spaces of higher dimensions. Because of the the geometrical interpretation, involving movement of vectors, such a transformation is often called a **linear mapping**.

30

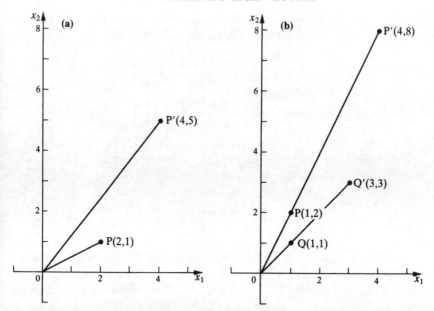

Figure 1.2 Linear mappings of vectors in two dimensions: (a) where there is change of length and orientation; (b) where there is change of length but not of orientation.

Vector magnification

The example shown above, and illustrated in Figure 1.2a, involves both a lengthening of the vector and a change of orientation. We now enquire into a situation where a mapping produces a magnification of a vector but no change of direction. In two dimensions, a vector

$$\begin{bmatrix} x_1 \\ x_2 \end{bmatrix}$$

is changed to the vector

$$\lambda \begin{bmatrix} x_1 \\ x_2 \end{bmatrix}$$

where λ is a scalar – the magnification factor. So the transformation is written in matrix notation as

$$\mathbf{A}\mathbf{x} = \lambda\mathbf{x} \qquad (1.7)$$

The question now arises as to how 'free' is the choice of combinations of **A**, **x**, and λ to produce the régime of equation (1.7). We shall set out the algebra of the situation below in two vertical columns: on the left in general matrix notation and on the right the various matrices and vectors are

31

written out explicitly for the two-dimensional case. Notes about the process are given afterwards. We start with equation (1.7).

Step 1 $\mathbf{Ax} = \lambda\mathbf{x}$ $\begin{bmatrix} a_{11} & a_{12} \\ a_{21} & a_{22} \end{bmatrix}\begin{bmatrix} x_1 \\ x_2 \end{bmatrix} = \lambda\begin{bmatrix} x_1 \\ x_2 \end{bmatrix}$

Step 2 $\mathbf{Ax} - \lambda\mathbf{x} = 0$ $\begin{bmatrix} a_{11} & a_{12} \\ a_{21} & a_{22} \end{bmatrix}\begin{bmatrix} x_1 \\ x_2 \end{bmatrix} - \lambda\begin{bmatrix} x_1 \\ x_2 \end{bmatrix} = \begin{bmatrix} 0 \\ 0 \end{bmatrix}$

Step 3 $[\mathbf{A} - \lambda\mathbf{I}]\mathbf{x} = 0$ $\left[\begin{bmatrix} a_{11} & a_{12} \\ a_{21} & a_{22} \end{bmatrix} - \lambda\begin{bmatrix} 1 & 0 \\ 0 & 1 \end{bmatrix}\right]\begin{bmatrix} x_1 \\ x_2 \end{bmatrix} = \begin{bmatrix} 0 \\ 0 \end{bmatrix}$

or

Step 3a $\begin{bmatrix} a_{11} - \lambda & a_{12} \\ a_{21} & a_{22} - \lambda \end{bmatrix}\begin{bmatrix} x_1 \\ x_2 \end{bmatrix} = \begin{bmatrix} 0 \\ 0 \end{bmatrix}$

Step 2 involves subtracting $\lambda\mathbf{x}$ from both sides of (1.7). Step 3 factorises out the column vector $\mathbf{x}$, and so the unit matrix has to be invoked in order to preserve the conformability for subtraction. The pair of linear equations shown at Step 3a is obviously homogeneous, and for $\mathbf{x}$ to have a non-trivial solution ($\mathbf{x} \neq 0$) the system must be degenerate (page 23). This means that the matrix $\mathbf{A} - \lambda\mathbf{I}$ must be singular, which further implies that its determinant must be zero. Thus we have

$$\begin{vmatrix} a_{11} - \lambda & a_{12} \\ a_{21} & a_{22} - \lambda \end{vmatrix} = (a_{11} - \lambda)(a_{22} - \lambda) - a_{21}a_{12} = 0 \qquad (1.8)$$

i.e.

$$a_{11}a_{22} - a_{11}\lambda - a_{22}\lambda + \lambda^2 - a_{21}a_{12} = 0$$

or

$$\lambda^2 - (a_{11} + a_{22})\lambda + (a_{11}a_{22} - a_{21}a_{12}) = 0$$

This is a quadratic equation in λ whose coefficients comprise only the elements of matrix $\mathbf{A}$. Hence values of λ are specific to a particular matrix: in the present two-dimensional case we find λ given as the roots of a quadratic equation and these will typically be two in number. Thus it seems that no more than two real values of λ can be associated with any particular matrix of order 2.

Once the λs have been evaluated, the degenerate homogeneous system of Step 3a above,

$$\left. \begin{array}{r} (a_{11} - \lambda)x_1 + a_{12}x_2 = 0 \\ a_{21}x_1 + (a_{22} - \lambda)x_2 = 0 \end{array} \right\} \qquad (1.9)$$

can be solved, using each value of λ in turn. This yields

$$x_2 = \frac{\lambda - a_{11}}{a_{12}} x_1$$

or in vector form

$$\begin{bmatrix} x_1 \\ x_2 \end{bmatrix} = \mu \begin{bmatrix} 1 \\ \dfrac{\lambda - a_{11}}{a_{12}} \end{bmatrix}$$

where μ is an arbitrary scalar which can take any value. Thus the nature of the vector that is magnified is entirely dependent on λ and the elements of matrix **A**.

The magnification factors, λ, are known variously as the **eigenvalues**, **latent roots**, and **characteristic roots** of the matrix **A**. The corresponding vectors, **x**, are alled **eigenvectors**, **latent vectors**, or **characteristic vectors** of **A**; and (1.8) is called the **characteristic equation**.

To fix ideas, pursue the following example, which also extends this topic to the three-dimensional case.

Example 1.14   Find the eigenvalues and eigenvectors of the following matrices, and draw the geometrical interpretation.

$$(a) \begin{bmatrix} 2 & 1 \\ -2 & 5 \end{bmatrix} \qquad (b) \begin{bmatrix} 1 & 3 & 0 \\ 7 & 1 & 5 \\ 0 & -4 & 1 \end{bmatrix}$$

(a)   The characteristic equation is

$$\begin{vmatrix} 2 - \lambda & 1 \\ -2 & 5 - \lambda \end{vmatrix} = (2 - \lambda)(5 - \lambda) + 2 = 0$$

i.e.

$$\lambda^2 - 7\lambda + 12 = 0$$

There is no need to use the usual formula to solve this equation since it factorises easily (two numbers that add to give -7 and multiply to give 12, i.e. -4 and -3). Hence

$$(\lambda - 4)(\lambda - 3) = 0$$

and so either $\lambda = 4$ or $\lambda = 3$. For $\lambda = 4$, system (1.9) becomes

$$(2 - 4)x_1 + \qquad x_2 = 0$$

$$-2x_1 + (5 - 4)x_2 = 0$$

i.e.

$$-2x_1 + x_2 = 0$$

$$-2x_1 + x_2 = 0$$

and so $x_2 = 2x_1$. In vector form

$$\begin{bmatrix} x_1 \\ x_2 \end{bmatrix} = \mu \begin{bmatrix} 1 \\ 2 \end{bmatrix}$$

In particular, if $\mu = 1$, then the eigenvector $\begin{bmatrix} 1 \\ 2 \end{bmatrix}$ corresponds to the eigenvalue $\lambda = 4$. The vector $\overrightarrow{OP}$ becomes the vector $\overrightarrow{OP'}$ in Figure 1.2b. For $\lambda = 3$, system (1.9) becomes

$$(2-3)x_1 + \qquad x_2 = 0$$

$$-2x_1 + (5-3)x_2 = 0$$

i.e.

$$-x_1 + x_2 = 0$$

$$-2x_1 + 2x_2 = 0$$

and so $x_1 = x_2$. In vector form

$$\begin{bmatrix} x_1 \\ x_2 \end{bmatrix} = \mu \begin{bmatrix} 1 \\ 1 \end{bmatrix}$$

In particular, if $\mu = 1$, then the eigenvector $\begin{bmatrix} 1 \\ 1 \end{bmatrix}$ corresponds to the eigenvalue $\lambda = 3$. The vector $\overrightarrow{OQ}$ becomes the vector $\overrightarrow{OQ'}$ in Figure 1.2b.

(b) The characteristic equation is

$$\begin{vmatrix} 1-\lambda & 3 & 0 \\ 7 & 1-\lambda & 5 \\ 0 & -4 & 1-\lambda \end{vmatrix} = 0$$

i.e.

$$(1-\lambda)^3 + (3)(5)(0) + (0)(7)(-4) - (0)(1-\lambda)(0)$$
$$- (-4)(5)(1-\lambda) - (1-\lambda)(7)(3) = 0$$

by the Rule of Sarrus; or

$$(1-\lambda)^3 + 20(1-\lambda) - 21(1-\lambda) = 0$$

34

i.e.

$$(1 - \lambda)^3 - (1 - \lambda) = 0$$

Expanding,

$$(1 - 3\lambda + 3\lambda^2 - \lambda^3) - (1 - \lambda) = 0$$

and, removing all brackets, we get

$$-\lambda^3 + 3\lambda^2 - 2\lambda = 0$$

For convenience change signs, that is, multiply both sides by -1:

$$\lambda^3 - 3\lambda^2 + 2\lambda = 0$$

Factorising: first

$$\lambda(\lambda^2 - 3\lambda + 2) = 0$$

then (two numbers which add to give -3 and multiply to give 2: -2 and -1)

$$\lambda(\lambda - 1)(\lambda - 2) = 0$$

So either $\lambda = 0$ (eigenvector reduced to a point at the origin), or $\lambda = 1$ (eigenvector remains unchanged), or $\lambda = 2$ (eigenvector doubled in length). For $\lambda = 2$, the three-dimensional analogue of equation system (1.9) is

$$(1 - 2)x_1 + \quad 3x_2 \quad\quad\quad = 0$$
$$7x_1 + (1 - 2)x_2 + \quad\quad 5x_3 = 0$$
$$-4x_2 + \quad (1 - 2)x_3 = 0$$

i.e.

$$-x_1 + 3x_2 \quad\quad = 0$$
$$7x_1 - \quad x_2 + 5x_3 = 0$$
$$-4x_2 - \quad x_3 = 0$$

Solving in the usual way gives

$$\begin{bmatrix} x_1 \\ x_2 \\ x_3 \end{bmatrix} = \mu \begin{bmatrix} 3 \\ 1 \\ -4 \end{bmatrix}$$

In particular, if $\mu = 1$, then the eigenvector

$$\begin{bmatrix} 3 \\ 1 \\ -4 \end{bmatrix}$$

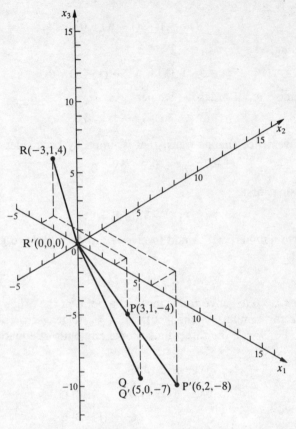

Figure 1.3 Linear mappings of vectors in three dimensions where there is change of length but not of orientation.

corresponds to the eigenvalue $\lambda = 2$. The vector $\overrightarrow{OP}$ becomes the vector $\overrightarrow{OP'}$ in Figure 1.3.

For $\lambda = 1$, the system equivalent to (1.9) is

$$(1-1)x_1 + \qquad 3x_2 \qquad\qquad = 0$$

$$7x_1 + (1-1)x_2 + \qquad 5x_3 = 0$$

$$-4x_2 + (1-1)x_3 = 0$$

i.e.

$$3x_2 \qquad\qquad = 0$$

$$7x_1 \qquad\qquad + \qquad 5x_3 = 0$$

$$-4x_2 \qquad\qquad = 0$$

36

from which

$$\begin{bmatrix} x_1 \\ x_2 \\ x_3 \end{bmatrix} = \mu \begin{bmatrix} 5 \\ 0 \\ -7 \end{bmatrix}$$

So, with $\mu = 1$ the vector $\overrightarrow{OQ}$ becomes $\overrightarrow{OQ'}$ (in Fig. 1.3), which is the same length, as the eigenvalue is unity.

Finally, for $\lambda = 0$, the system analagous to (1.9) is

$$x_1 + 3x_2 \qquad = 0$$
$$7x_1 + \quad x_2 + 5x_3 = 0$$
$$- 4x_2 + \quad x_3 = 0$$

which gives

$$\begin{bmatrix} x_1 \\ x_2 \\ x_3 \end{bmatrix} = \mu \begin{bmatrix} -3 \\ 1 \\ 4 \end{bmatrix}$$

Because the eigenvalue is zero, any eigenvector is shrunk to zero in the transformation. For any value of μ, vector $\overrightarrow{OR}$ becomes $\overrightarrow{OR'}$ (in Fig. 1.3), which is simply a point at the origin.

Reduction of a matrix to diagonal form

If we have a square matrix $\mathbf{A}$ and we find its eigenvalues and eigenvectors, then we may convert $\mathbf{A}$ to a diagonal matrix. We do this by forming a new square matrix, $\mathbf{Q}$, whose columns are the eigenvectors of $\mathbf{A}$; then

$$\mathbf{Q}^{-1}\mathbf{AQ} = \mathbf{\Lambda} \qquad (1.10)$$

in which $\mathbf{\Lambda}$ is a diagonal matrix whose elements are the eigenvalues of $\mathbf{A}$. Since $\mathbf{\Lambda}$ is derived solely from $\mathbf{A}$, we can regard $\mathbf{\Lambda}$ as the diagonal form of A.

Example 1.15   Find the diagonal forms of

(a) $\begin{bmatrix} 2 & 1 \\ -2 & 5 \end{bmatrix}$ (b) $\begin{bmatrix} 1 & 3 & 0 \\ 7 & 1 & 5 \\ 0 & -4 & 1 \end{bmatrix}$

These are the same matrices whose eigenvalues and eigenvectors have already been found in Example 1.14.

37

(a) We already have $\lambda_1 = 4$ and $\lambda_2 = 3$, giving corresponding eigenvectors of

$\begin{bmatrix} 1 \\ 2 \end{bmatrix}$ and $\begin{bmatrix} 1 \\ 1 \end{bmatrix}$. Thus we have

$$Q = \begin{bmatrix} 1 & 1 \\ 2 & 1 \end{bmatrix} \qquad |Q| = -1$$

$$Q^{-1} = \begin{bmatrix} -1 & 1 \\ 2 & -1 \end{bmatrix}$$

$$AQ = \begin{bmatrix} 2 & 1 \\ -2 & 5 \end{bmatrix}\begin{bmatrix} 1 & 1 \\ 2 & 1 \end{bmatrix} = \begin{bmatrix} 4 & 3 \\ 8 & 3 \end{bmatrix}$$

$$Q^{-1}AQ = \begin{bmatrix} -1 & 1 \\ 2 & -1 \end{bmatrix}\begin{bmatrix} 4 & 3 \\ 8 & 3 \end{bmatrix} = \begin{bmatrix} 4 & 0 \\ 0 & 3 \end{bmatrix} = \Lambda$$

Hence the matrix $\begin{bmatrix} 4 & 0 \\ 0 & 3 \end{bmatrix}$ is the diagonal form of $\begin{bmatrix} 2 & 1 \\ -2 & 5 \end{bmatrix}$. Notice that

$$|A| = |\Lambda|$$

The determinant of the diagonal form of a matrix is the same as the matrix itself.

(b) We already have $\lambda_1 = 2$, $\lambda_2 = 1$, $\lambda_3 = 0$, with corresponding eigenvectors of

$$\begin{bmatrix} 3 \\ 1 \\ -4 \end{bmatrix}, \qquad \begin{bmatrix} 5 \\ 0 \\ -7 \end{bmatrix} \quad \text{and} \quad \begin{bmatrix} -3 \\ 1 \\ 4 \end{bmatrix}$$

Thus we have

$$Q = \begin{bmatrix} 3 & 5 & -3 \\ 1 & 0 & 1 \\ -4 & -7 & 4 \end{bmatrix} \quad \text{and} \quad |Q| = -20 + 21 + 21 - 20 = 2$$

by the Rule of Sarrus. Then

$$\text{adj } Q = \begin{bmatrix} 7 & 1 & 5 \\ -8 & 0 & -6 \\ -7 & 1 & -5 \end{bmatrix} \quad \text{and} \quad Q^{-1} = \tfrac{1}{2}\begin{bmatrix} 7 & 1 & 5 \\ -8 & 0 & -6 \\ -7 & 1 & -5 \end{bmatrix}$$

and so

$$AQ = \begin{bmatrix} 1 & 3 & 0 \\ 7 & 1 & 5 \\ 0 & -4 & 1 \end{bmatrix}\begin{bmatrix} 3 & 5 & -3 \\ 1 & 0 & 1 \\ -4 & -7 & 4 \end{bmatrix} = \begin{bmatrix} 6 & 5 & 0 \\ 2 & 0 & 0 \\ -8 & -7 & 0 \end{bmatrix}$$

and

$$Q^{-1}AQ = \tfrac{1}{2} \begin{bmatrix} 7 & 1 & 5 \\ -8 & 0 & -6 \\ -7 & 1 & -5 \end{bmatrix} \begin{bmatrix} 6 & 5 & 0 \\ 2 & 0 & 0 \\ -8 & -7 & 0 \end{bmatrix} = \tfrac{1}{2} \begin{bmatrix} 4 & 0 & 0 \\ 0 & 2 & 0 \\ 0 & 0 & 0 \end{bmatrix}$$

$$= \begin{bmatrix} 2 & 0 & 0 \\ 0 & 1 & 0 \\ 0 & 0 & 0 \end{bmatrix} = \Lambda$$

The rank of a matrix

Linear dependence of vectors

A matrix may be regarded as consisting of a set of column vectors or a set of row vectors. For example, the matrix

$$\begin{bmatrix} 2 & 5 & 2 \\ 3 & 1 & 7 \\ -4 & -3 & 4 \end{bmatrix}$$

can be thought of as the set of column vectors

$$\begin{bmatrix} 2 \\ 3 \\ -4 \end{bmatrix}, \quad \begin{bmatrix} 5 \\ 1 \\ -3 \end{bmatrix} \quad \text{and} \quad \begin{bmatrix} 2 \\ 7 \\ 4 \end{bmatrix}$$

or the set of row vectors

$$[2 \quad 5 \quad 2], \quad [3 \quad 1 \quad 7], \quad \text{and} \quad [-4 \quad -3 \quad 4]$$

A linear combination of the vectors $a_1, a_2, \ldots, a_n$ is defined to be

$$k_1 a_1 + k_2 a_2 + \ldots + k_n a_n$$

where $k_1, k_2, \ldots, k_n$ is a set of scalars. Each of the vectors $\{a_i\}$ must contain the same number of elements. If for a set of vectors $a_1, a_2, \ldots, a_n$ a set of scalars $k_1, k_2, \ldots, k_n$ can be found such that

$$k_1 a_1 + k_2 a_2 + \ldots + k_n a_n = 0$$

where 0 is a corresponding sized vector of zeros, then the vectors are said to be **linearly dependent**; otherwise the vectors are **linearly independent**. We exclude the cases where all $a_i = 0$ and/or all $k_i = 0$. For either the set of column vectors or row vectors of the above matrix example, no such set of scalars exists; hence the above vectors are linearly independent.

On the other hand, the matrix

$$\begin{bmatrix} 2 & 3 & -2 \\ 1 & -2 & 3 \\ 4 & -1 & 4 \end{bmatrix}$$

can be decomposed into the set of column vectors,

$$\begin{bmatrix} 2 \\ 1 \\ 4 \end{bmatrix}, \quad \begin{bmatrix} 3 \\ -2 \\ -1 \end{bmatrix} \quad \text{and} \quad \begin{bmatrix} -2 \\ 3 \\ 4 \end{bmatrix}$$

and it can be shown that

$$-5\begin{bmatrix} 2 \\ 1 \\ 4 \end{bmatrix} + 8\begin{bmatrix} 3 \\ -2 \\ -1 \end{bmatrix} + 7\begin{bmatrix} -2 \\ 3 \\ 4 \end{bmatrix} = \begin{bmatrix} 0 \\ 0 \\ 0 \end{bmatrix}$$

Likewise, this matrix can be partitioned into a set of row vectors

$$[2 \quad 3 \quad -2], [1 \quad -2 \quad 3], \text{ and } [4 \quad -1 \quad 4],$$

and it can also be shown that

$$-1[2 \quad 3 \quad -2] - 2[1 \quad -2 \quad 3] + 1[4 \quad -1 \quad 4] = [0 \quad 0 \quad 0]$$

So, for the second example matrix, the vectors are linearly dependent. Note that if we remove just one row or column the remaining row or column vectors are linearly independent; but for the matrix

$$\begin{bmatrix} 6 & 3 & 12 \\ 8 & 4 & 16 \\ 12 & 6 & 24 \end{bmatrix}$$

in which the vectors are linearly dependent, removal of any one row or column does not affect the situation – the remaining two rows or columns are still linearly dependent.

The rank of a matrix

The rank of a matrix is a most important concept, and we start with the definition.

The rank of a matrix is the number of linearly independent rows (or columns) in the matrix.

The rank of a matrix, **A**, is usually denoted as $r(\mathbf{A})$. Only the zero, or null, matrix has zero rank; for any other matrix the r-value is a positive integer.

It is convenient to consider rank in relation to square and rectangular matrices separately.

SQUARE MATRICES

Here, there are two situations to discuss.

(1) If the number of linearly independent rows (or columns) is equal to the total number of rows (or columns), the matrix is said to be of **full rank**. The determinant of the matrix is not zero, that is, the matrix is not singular and its inverse therefore exists. Linear equations whose co-efficients form a square matrix of full rank are a non-degenerative system.

(2) If the number of linearly independent rows (or columns) is less than the total number of rows (or columns), the matrix is not of full rank. The determinant of the matrix is zero, that is, the matrix is singular and its inverse does not exist. Linear equations whose coefficients form a square matrix which is not of full rank are a degenerative system.

RECTANGULAR MATRICES

In an ($m \times n$) matrix, the rank cannot be greater than the lesser of m and n. Thus if $m < n$, the matrix has more columns than rows, and the columns are linearly dependent. This is true even if the rows are linearly independent (full row rank).

Determining the rank of a matrix

As for most matrix operations, computer routines are available to determine the rank of a matrix, and again, these routines are usually based on the elementary transformations. The procedures will be demonstrated below for some small matrix examples.

Example 1.16 Determine the rank of each of the following matrices:

(a) $\begin{bmatrix} 2 & 5 & 2 \\ 3 & 1 & 7 \\ -4 & -3 & 4 \end{bmatrix}$ 　(b) $\begin{bmatrix} 2 & 3 & -2 \\ 1 & -2 & 3 \\ 4 & -1 & 4 \end{bmatrix}$ 　(c) $\begin{bmatrix} 6 & -4 & 10 \\ -3 & 2 & -5 \\ 12 & -8 & 20 \end{bmatrix}$

(d) $\begin{bmatrix} 1 & 4 & 6 & 0 \\ 4 & -1 & 2 & 0 \\ -7 & 6 & 2 & 1 \end{bmatrix}$ 　(e) $\begin{bmatrix} 4 & -1 \\ 2 & 3 \\ 5 & 9 \end{bmatrix}$

(a) First write out the matrix without brackets, and label the rows:

$$\begin{array}{ccc} 2 & 5 & 2 \end{array} \qquad (1)$$

$$\begin{array}{ccc} 3 & 1 & 7 \end{array} \qquad (2)$$

$$\begin{array}{ccc} -4 & -3 & 4 \end{array} \qquad (3)$$

This matrix is the same as the matrix of coefficients of the set of linear equations in Example 1.9a, so the same operations are carried out, resulting in

$$\begin{array}{ccc} 1 & \tfrac{5}{2} & 1 \end{array} \qquad (1a)$$

$$\begin{array}{ccc} 0 & 1 & -\tfrac{8}{13} \end{array} \qquad (2b)$$

$$\begin{array}{ccc} 0 & 0 & \tfrac{160}{3} \end{array} \qquad (3b)$$

(compare with page 20). Here we end up with three non-zero row vectors, the same as the order of the matrix, so the matrix is of full rank.

(b) This is the same matrix as that of the coefficients of the set of linear equations in Example 1.10; so starting with the labelled rows

$$\begin{array}{ccc} 2 & 3 & -2 \end{array} \qquad (1)$$

$$\begin{array}{ccc} 1 & -2 & 3 \end{array} \qquad (2)$$

$$\begin{array}{ccc} 4 & -1 & 4 \end{array} \qquad (3)$$

and following through the transformations as on page 23, we arrive at

$$\begin{array}{ccc} 1 & -2 & 3 \end{array} \qquad (2)$$

$$\begin{array}{ccc} 0 & 7 & -8 \end{array} \qquad (1a)$$

$$\begin{array}{ccc} 0 & 0 & 0 \end{array} \qquad (3b)$$

There are only two non-zero row vectors, so the rank is 2.

(c) Starting as before:

$$\begin{array}{ccc} 6 & -4 & 10 \end{array} \qquad (1)$$

$$\begin{array}{ccc} -3 & 2 & -5 \end{array} \qquad (2)$$

$$\begin{array}{ccc} 12 & -8 & 20 \end{array} \qquad (3)$$

Divide (1) by 2:

$$\begin{array}{ccc} 3 & -2 & 5 \end{array} \qquad (1a)$$

Add (1a) to (2) and write out the new system:

$$\begin{array}{ccc} 6 & -4 & 10 \end{array} \qquad (1)$$

$$\begin{array}{ccc} 0 & 0 & 0 \end{array} \qquad (2a)$$

$$\begin{array}{ccc} 12 & -8 & 20 \end{array} \qquad (3)$$

Multiply *(1)* by -2:

$$-12 \qquad 8 \qquad -20 \qquad\qquad (1b)$$

Add *(1b)* to *(3)* and write out the final system:

6	-4	10	*(1)*
0	0	0	*(2a)*
0	0	0	*(3a)*

As there is only one non-zero row, the rank is 1.

(d) We have

1	4	6	0	*(1)*
4	-1	2	0	*(2)*
-7	6	2	1	*(3)*

Multiplying *(1)* by -4, and adding the result to *(2)* gives the system

1	4	6	0	*(1)*
0	-17	-22	0	*(2a)*
-7	6	2	1	*(3)*

Multiplying *(1)* by 7, and adding the result to *(3)* gives

1	4	6	0	*(1)*
0	-17	-22	0	*(2a)*
0	34	44	1	*(3a)*

Multiplying *(2a)* by 2, and adding the result to *(3a)* gives

1	4	6	0	*(1)*
0	-17	-22	0	*(2a)*
0	0	0	1	*(3b)*

This process cannot be taken further, and as we have three non-zero row vectors the rank is 3, that is, the matrix has full row rank.

(e) We have

4	-1	*(1)*
2	3	*(2)*
5	9	*(3)*

Multiplying *(1)* by $-\frac{1}{2}$, and adding the result to *(2)* gives

4	-1	*(1)*
0	$\frac{7}{2}$	*(2a)*
5	9	*(3)*

Multiplying *(1)* by $-\frac{5}{4}$ and adding the result to *(3)* gives

4	-1	*(1)*
0	$\frac{7}{2}$	*(2a)*
0	$\frac{41}{4}$	*(3a)*

Divide *(2a)* by $-\frac{7}{2}$, multiply the result by $\frac{41}{4}$, and add the final result to *(3a)*, giving

4	-1	*(1)*
0	$\frac{7}{2}$	*(2a)*
0	0	*(3b)*

As we have only two non-zero row vectors, the matrix has rank 2, i.e. full column rank.

Determining the coefficients of linearly dependent vectors

If we know that a matrix is less than full rank, it may be necessary to know the coefficients $\{k_i\}$ of the linearly dependent vectors (see page 39). One simple example will suffice to demonstrate the procedure.

Example 1.17   Find the coefficient values of the linearly dependent column vectors of the matrix

$$\begin{bmatrix} 1 & -3 & 2 \\ 3 & 0 & -1 \end{bmatrix}$$

Because there are more columns than rows, then the former are linearly dependent vectors. It is also easy to show that the rank of this matrix is 2. We have

$$k_1\begin{bmatrix} 1 \\ 3 \end{bmatrix} + k_2\begin{bmatrix} -3 \\ 0 \end{bmatrix} + k_3\begin{bmatrix} 2 \\ -1 \end{bmatrix} = \begin{bmatrix} 0 \\ 0 \end{bmatrix}$$

Expansion gives the pair of linear equations

$$k_1 - 3k_2 + 2k_3 = 0$$
$$3k_1 \qquad - k_3 = 0$$

from which $k_3 = 3k_1$ and $k_2 = \frac{7}{3}k_1$. So

$$\begin{bmatrix} k_1 \\ k_2 \\ k_3 \end{bmatrix} = \lambda \begin{bmatrix} 1 \\ \frac{7}{3} \\ 3 \end{bmatrix} = \mu \begin{bmatrix} 3 \\ 7 \\ 9 \end{bmatrix}$$

where $\mu = 3\lambda$. Any constant multiplier, e.g. λ or μ, is irrelevant; it is the ratios between the k_i that are relevant here.

Suggestions for further reading

As a source of required elementary material for reading the present book, the following is recommended: **Causton, D. R.** (1983). *A biologist's basic mathematics*. London: Edward Arnold.

There are numerous books on matrices and linear algebra, as well as chapters on these topics in many mathematics textbooks. However, I have found the following particularly useful.

If you have found this chapter hard going, and are still rather confused, try **Coulson, A. E.** (1965). *Introduction to matrices*. London: Longman. This book is extremely simply written, with worked examples in which every step is explicitly given.

Another small book which deals thoroughly with linear equations, plus an introduction to matrices, is **Cohn, P. M.** (1965). *Linear equations* (Library of mathematics series). London: Routledge & Kegan Paul.

An unusual book, which is adequately described by its title but is both rigorous and full of interest, is **Fletcher, T. J.** (1972). *Linear algebra through its applications*. New York: Van Nostrand Reinhold.

Two related books, the first of which is of obvious interest to the biologist, are **Searle, S. R.** (1966). *Matrix algebra for the biological sciences*. New York: Wiley and **Searle, S. R.** and **W. H. Hausman** (1970). *Matrix algebra for business and economics*. New York: Wiley.

A further reference quoted in this chapter is **Nicolson, M. M., D. R. Hartree,** and **D. G. Padfield** (1961). *Fundamentals and techniques of mathematics for scientists*. London: Longman.

Exercises

1.1 Evaluate

(a) $\begin{vmatrix} 4 & -6 \\ 2 & 5 \end{vmatrix}$

(b) $\begin{vmatrix} 1 & 0 & 2 \\ 3 & 4 & 5 \\ 5 & 6 & 7 \end{vmatrix}$

LINEAR ALGEBRA

1.2 Evaluate

(a) $\begin{vmatrix} 2 & 5 & 2 \\ 3 & 1 & 7 \\ -4 & -3 & 4 \end{vmatrix}$

(b) $\begin{vmatrix} 2 & 1 & 2 & -1 \\ 6 & 8 & 12 & -13 \\ 10 & 2 & 2 & 3 \\ -4 & 4 & 1 & -3 \end{vmatrix}$

1.3 Let

$$A = \begin{bmatrix} 1 & 2 & 1 \\ 2 & 4 & 6 \\ 3 & 1 & 2 \end{bmatrix} \qquad B = \begin{bmatrix} -1 & 2 & -3 \\ 2 & -1 & 4 \\ 3 & 4 & 1 \end{bmatrix}$$

Find (a) **AB**   (b) **BA**   (c) $|A|$

      (d) $|B|$ (e) $|AB|$ (f) $|BA|$

What do these results show?

1.4 Find the adjoint matrices of **A** and **B** in Exercise 1.3 above. Hence, knowing the determinants, write down the matrices A^{-1} and B^{-1}.

1.5 Evaluate $|A^{-1}|$ and $|B^{-1}|$ of the two previous exercises. What do these results show?

1.6 Solve the following systems of linear equations:

(a) $3x_1 - 3x_2 + 5x_3 = 6$
$x_1 + 7x_2 + 5x_3 = 4$
$5x_1 + 10x_2 + 15x_3 = 9$

(b) $4x_1 - 8x_2 + 4x_3 - 20x_4 = 0$
$3x_1 - 6x_2 + 4x_3 - 14x_4 = 4$
$-2x_1 + 4x_2 - 4x_3 + 9x_4 = 5$
$4x_1 - 7x_2 + 5x_3 - 8x_4 = 6$

(c) $x_1 + x_2 + x_3 = 24$
$200x_1 - 50x_2 - 150x_3 = 0$

(d) $2x_1 + 4x_2 + x_3 = 1$
$3x_1 + 5x_2 = 1$
$5x_1 + 13x_2 + 7x_3 = 4$

(e) $9x_1 - 6x_2 + 12x_3 = 0$
$-12x_1 + 8x_2 - 16x_3 = 0$
$-7x_1 + 10x_2 - 13x_3 = 0$

EXERCISES

(f) $2x_1 + 3x_2 + 4x_3 = 1$

$5x_1 + 6x_2 + 7x_3 = 2$

$8x_1 + 9x_2 + 10x_3 = 4$

(g) $x_1 + x_2 + x_3 = 2$

$x_1 + 2x_2 + 3x_3 = 3$

$x_1 + 3x_2 + 5x_3 = 4$

1.7 Find the eigenvalues and eigenvectors, and write down the diagonal form, of each of the following matrices:

(a) $\begin{bmatrix} 2 & 4 \\ 3 & 13 \end{bmatrix}$ (b) $\begin{bmatrix} 1 & -1 & -2 \\ 2 & 4 & 2 \\ 1 & 1 & 4 \end{bmatrix}$

1.8 Find the ranks of the matrices of coefficients of the linear equations in Exercise 1.6.

1.9 In Exercise 1.8, where the matrices have less than full rank find the coefficients of the linearly dependent vectors.

2

Applications of linear algebra

The summarisation of multivariate data

In this section is will be assumed that the reader has a knowledge of basic statistical theory and methods for univariate situations, that is, situations where only one biological (or perhaps, environmental) quantity is measured. Very often it is found that if a large number of replicate measurements are made the distribution of the observations approximates closely to the normal distribution. This distribution is usually described by its **probability density curve,** whose equation is

$$f(x) = \frac{1}{\sigma\sqrt{(2\pi)}} \exp\left\{ -\tfrac{1}{2}\left(\frac{x-\mu}{\sigma}\right)^2\right\} \tag{2.1}$$

where x is the measurement, and μ and σ are constants. The variable, x, because of its nature, is known as a **random variable** or **variate;** μ is the mean, which specifies the position of the centre of the distribution of x; and σ is the standard deviation (σ^2 is the variance), which quantifies the spread of the curve about the mean and so the variability of the **statistical population** of measurements involved. Equation (2.1) represents the complete family of normal distributions; when we have particular numerical values of μ and σ (provided as sample statistics, see below), we know which one of the family we need for a particular set of data. The quantities μ and σ are known as **population parameters.**

The concept of a statistical population is based on the idea of an infinity of similar measurements; consequently, a population is a theoretical, although extremely useful, entity and the values of its parameters for any particular set of measurements can never be actually known. In practice, any collection – large or small – of similar measurements is regarded as a sample drawn from the population. The summary statistics which may be calculated from the data, e.g. mean, standard deviation, and variance, are not the population parameters but are only *estimates* of them. For this reason one must distinguish between the population parameters, whose

48

exact value can never be known but only inferred or hypothesised, and **sample statistics,** which are calculated from a sample of data and are estimates of the corresponding population parameters. The symbols given to the sample statistics are $\bar{x}$ for the mean, s for the standard deviation, and s^2 for the variance. Notice the convention of using greek letters for population parameters and roman letters for sample statistics.

There are many situations in which more than one kind of measurement needs to be made within the system. For example, in the quantitative study of morphology (morphometrics) a number of different features on each of the number of replicate organisms may need to be measured. This gives rise to multivariate statistical theory and methods, and there are multivariate analogues of the univariate normal distribution. Summarising multivariate data implies calculating sample statistics. Exactly how such a summary should be presented depends to some extent on the data themselves and the use to which they are to be put; a universally acceptable way, which forms a starting point for the various multivariate analyses available, is to write the summaries in matrix form. We shall start with the simplest case where there are only two variates involved – **bivariate** populations and samples.

Bivariate population

The bivariate analogue of the normal distribution is the **bivariate normal distribution.** It has five parameters. Denoting the two variates as 1 and 2, the parameters are $\mu_1, \mu_2, \sigma_1, \sigma_2, \rho_{12}$, where μ_1 and μ_2 are the means of variates 1 and 2, σ_1 and σ_2 the corresponding standard deviations, and ρ_{12} is called the correlation coefficient. The first four parameters relate to each variate quite independently of the other, whereas the last parameter measures the degree of association between the two variates, that is, what degree of tendency there is for, say, large values of variate 1 to be associated with large values of variate 2.

The correlation coefficient is a particularly useful parameter in that it lies in the range $-1 \leqslant \rho_{12} \leqslant 1$. A value near 1 indicates that high values of one variate strongly tend to be associated with high values of the other and vice versa (positive correlation); a high negative correlation, near -1, indicates that high values of one variate strongly tend to be associated with low values of the other (negative correlation); and a near zero value shows there to be little or no association between different values of the two variates.

There is, however, another parameter that expresses the degree of association between the two variates, which is more fundamental but does not give the same 'at a glance' assessment as does the correlation coefficient. This other parameter is the **covariance** between the two variates and, although it may be positive or negative according to whether positive or negative correlation is present and zero for the no correlation situation, it

APPLICATIONS OF LINEAR ALGEBRA

does not have the limits to its range of ± 1. Consequently it is impossible to use the covariance to assess the degree of correlation.

The covariance between variates 1 and 2 is given by

$$\sigma_{12} = \rho_{12}\sigma_1\sigma_2 \tag{2.2}$$

By its name, the covariance is a related quantity to the variance. If instead of giving, say, the variance of variate 1 the symbol σ_1^2 we give it σ_{11} (the double subscript serves to distinguish it from the standard deviation σ_1), then variate 2 has variance σ_{22}, and the two variates have covariance σ_{12}, as on the left-hand side of (2.2).

Now we are in a position to summarise in matrix form the population parameters of the bivariate normal distribution. We have

$$\boldsymbol{\mu} = \begin{bmatrix} \mu_1 \\ \mu_2 \end{bmatrix} \qquad \boldsymbol{\Sigma} = \begin{bmatrix} \sigma_{11} & \sigma_{12} \\ \sigma_{21} & \sigma_{22} \end{bmatrix} \qquad \mathbf{P} = \begin{bmatrix} 1 & \rho_{12} \\ \rho_{21} & 1 \end{bmatrix}$$

The first matrix is a column vector, $\boldsymbol{\mu}$, and is called the **vector of means.** The second matrix contains the variances and covariance and is known as the **variance–covariance matrix.** There is, of course, no difference between σ_{21} and σ_{12}; both symbols represent the population covariance and so the matrix is not only square but is symmetrical. Notice that the leading diagonal contains the variances and that the other elements are the covariance. Adhering to the convention of giving the whole matrix the symbol of a capital letter in heavy type corresponding to the small letter used to denote the elements inside the matrix, we have, in this instance, the capital greek letter 'sigma', $\boldsymbol{\Sigma}$, as shown. There is the possibility of confusing this use of $\boldsymbol{\Sigma}$ with that of $\sum$ which is the summation sign. The danger of confusion should be minimised if it is always borne in mind that a symbol in heavy type denotes a matrix.

The third matrix, $\mathbf{P}$, is closely similar to the variance–covariance matrix; it contains correlation coefficients and is called the **correlation matrix.** Even the units in the leading diagonal are correlation coefficients because each represents the correlation of one variate with itself, which must be perfect and positive, and hence unity. The off-diagonal elements are both the correlation coefficient of variate 1 and variate 2, and so again this matrix is symmetric. The capital greek letter 'rho' is indistinguishable from the roman P.

Another way of viewing the correlation matrix is as the variance–covariance matrix of **standardised variates.** If every variate in each observation is divided by its standard deviation and the variance–covariance matrix of these standardised variates is obtained, it is precisely the correlation matrix. Evidently the variances of all standardised variates are the same – unity. For further details, see page 72.

Bivariate sample

For a sample taken from a bivariate normally distributed population, sample means ($\bar{x}_1$ and $\bar{x}_2$), sample variances (s_{11} and s_{22}), and the sample covariance and correlation coefficient (s_{21} and r_{21}) can be calculated. The formulae for doing these calculations are

$$\bar{x}_1 = \frac{1}{n} \sum x_1$$

$$s_{11} = \frac{1}{n-1} \sum (x_1 - \bar{x}_1)^2 = \frac{1}{n-1} \left\{ \sum x_1^2 - \frac{(\sum x_1)^2}{n} \right\}$$

with identical forms for variate 2. The formulae for the covariance and correlation coefficients are

$$s_{12} = s_{21} = \frac{1}{n-1} \sum (x_1 - \bar{x}_1)(x_2 - \bar{x}_2) = \frac{1}{n-1} \left\{ \sum x_1 x_2 - \frac{\sum x_1 \sum x_2}{n} \right\}$$

$$r_{12} = r_{21} = \frac{s_{21}}{\sqrt{(s_{11}s_{22})}}$$

where each summation sign indicates summation over all the replicate values for that variate. In the formulae for the variance and covariance, the expression to the right-hand side of the final equals sign is the easier to use when using a calculator. The results, which are clearly analogous to the population matrices, are put in the following matrix form:

$$\bar{x} = \begin{bmatrix} \bar{x}_1 \\ \bar{x}_2 \end{bmatrix} \qquad S = \begin{bmatrix} s_{11} & s_{12} \\ s_{21} & s_{22} \end{bmatrix} \qquad R = \begin{bmatrix} 1 & r_{12} \\ r_{21} & 1 \end{bmatrix}$$

where $\bar{x}$ is the sample vector of means, S is the sample variance–covariance matrix, and R is the sample correlation matrix.

The sample variance–covariance matrix can be derived in the following illuminating way. Bivariate data have to be displayed in a two-way table where, let us assume, the rows represent the replicate observations and the columns represent the variates. Let us take a very simple bivariate example in which there are only three observations. A two-way table can, of course, be written down as a matrix, and nowadays a multivariate data table is frequently referred to as a **data matrix.** Our present data matrix would thus appear as

$$X = \begin{bmatrix} x_{11} & x_{12} \\ x_{21} & x_{22} \\ x_{31} & x_{32} \end{bmatrix}$$

where the first subscript x_{ij} represents the replicate number, and the second subscript is the variate. So x_{ij} represents the value of the jth variate in the

51

ith observation. If we now subtract the mean of the jth variate from each x_{ij}, we have

$$\mathbf{X} = \begin{bmatrix} x_{11} - \bar{x}_1 & x_{12} - \bar{x}_2 \\ x_{21} - \bar{x}_1 & x_{22} - \bar{x}_2 \\ x_{31} - \bar{x}_1 & x_{32} - \bar{x}_2 \end{bmatrix}$$

and we retain the symbol $\mathbf{X}$ for simplicity. Next, pre-multiply $\mathbf{X}$ by its transpose, giving

$$\mathbf{X}^T\mathbf{X} = \begin{bmatrix} x_{11} - \bar{x}_1 & x_{21} - \bar{x}_1 & x_{31} - \bar{x}_1 \\ x_{12} - \bar{x}_2 & x_{22} - \bar{x}_2 & x_{32} - \bar{x}_2 \end{bmatrix} \begin{bmatrix} x_{11} - \bar{x}_1 & x_{12} - \bar{x}_2 \\ x_{21} - \bar{x}_1 & x_{22} - \bar{x}_2 \\ x_{31} - \bar{x}_1 & x_{32} - \bar{x}_2 \end{bmatrix}$$

$$= \begin{bmatrix} \sum (x_{i1} - \bar{x}_1)^2 & \sum (x_{i1} - \bar{x}_1)(x_{i2} - \bar{x}_2) \\ \sum (x_{i2} - \bar{x}_2)(x_{i1} - \bar{x}_1) & \sum (x_{i2} - \bar{x}_2)^2 \end{bmatrix}$$

On multiplying each element by the scalar $1/(n-1)$, the above matrix is precisely the variance–covariance matrix. Hence we have the following important relationship.

In a data matrix, $\mathbf{X}$, of n rows (observations) and m columns (variates), in which the elements are deviations of each measurement from its own variate mean, the variance–covariance matrix of the observations is given by

$$\mathbf{S} = \frac{1}{n-1} \mathbf{X}^T\mathbf{X} \tag{2.3}$$

Multivariate population and sample

The advantage of the matrix method of summarising multivariate data is that the same structure is used for a population consisting of any number of variates. Consider an m-variate population and sample.

Population

$$\boldsymbol{\mu} = \begin{bmatrix} \mu_1 \\ \mu_2 \\ \vdots \\ \mu_m \end{bmatrix}$$

$$\boldsymbol{\Sigma} = \begin{bmatrix} \sigma_{11} & & \\ \sigma_{21} & \sigma_{22} & \\ \vdots & \vdots & \\ \sigma_{m1} & \sigma_{m2} \ldots \sigma_{mm} \end{bmatrix}$$

Sample

$$\bar{\mathbf{x}} = \begin{bmatrix} \bar{x}_1 \\ \bar{x}_2 \\ \vdots \\ \bar{x}_m \end{bmatrix}$$

$$\mathbf{S} = \begin{bmatrix} s_{11} & & \\ s_{21} & s_{22} & \\ \vdots & \vdots & \\ s_{m1} & s_{m2} \ldots s_{mm} \end{bmatrix}$$

$$
\mathbf{P} = \begin{bmatrix} 1 & & \\ \rho_{21} & 1 & \\ \vdots & \vdots & \\ \rho_{m1} & \rho_{m2} & \cdots & 1 \end{bmatrix}
\qquad
\mathbf{R} = \begin{bmatrix} 1 & & \\ r_{21} & 1 & \\ \vdots & \vdots & \\ r_{m1} & r_{m2} & \cdots & 1 \end{bmatrix}
$$

The analogy with the bivariate situation is complete; the column vectors of means are of length m and the variance–covariance and correlation matrices are symmetric matrices of order m.

Example 2.1   The data in Table 2.1 are a subset (kept small for convenience) from a much larger collection on the morphology of a population (in the biological sense) of Silene maritima *(sea campion) plants. The characters (variates) are*

1 – length of a mature leaf in the middle of a stem;
2 – length of a fully grown pedicel;
3 – length of the internode beneath the measured leaf;
4 – length of a bract subtending the measured pedicel.

All measurements are in mm. Variates 1 and 3 were measured with a ruler to the nearest mm, while variates 2 and 4 were measured to the nearest 0.1 mm under a microscrope with a calibrated eyepiece.

The original purpose of the survey from which these data were extracted was to examine variation within and between populations of *Silene maritima* (J. P. Savidge, unpublished). This species consists of many

Table 2.1 A data matrix consisting of 10 observations and 4 variates. (For full explanation, see text.)

Observation	Variate 1	2	3	4
1	2.57	0.53	2.77	1.84
2	3.00	0.64	2.83	2.25
3	2.94	0.47	2.71	2.12
4	3.04	0.64	2.30	2.34
5	2.77	0.18	2.49	1.55
6	2.30	0.64	1.61	1.74
7	2.83	0.34	2.20	1.81
8	2.71	0.47	2.77	2.07
9	2.71	1.16	3.09	1.72
10	3.00	0.99	3.40	1.78

distinct populations (ecotypes), and the characteristics of these populations are of interest to ecological geneticists, particularly the evolution of heavy metal tolerance that has occurred in some of these populations.

Because the measurements are a result of growth, there are good reasons for logarithmically transforming the measurements before analysis (see Causton 1983, Ch. 9). Table 2.1, therefore, quotes the natural logarithm of each measurement correct to two decimal places. The matrix summaries of these data are

$$\bar{\mathbf{x}} = \begin{bmatrix} 2.787 \\ 0.606 \\ 2.617 \\ 1.922 \end{bmatrix} \quad \mathbf{S} = \begin{bmatrix} 0.0529 \\ 0.0045 & 0.083\,7 \\ 0.0599 & 0.071\,7 & 0.2491 \\ 0.0320 & 0.000\,02 & 0.0040 & 0.0661 \end{bmatrix}$$

$$\mathbf{R} = \begin{bmatrix} 1 \\ 0.0676 & 1 \\ 0.5218 & 0.4966 & 1 \\ 0.5412 & 0.0003 & 0.0312 & 1 \end{bmatrix}$$

All the correlations are positive, indicating that, in general, large plants tend to be large in all their parts; however, none of the correlation coefficients are particularly high, which implies that this tendency is fairly weak. In the variance–covariance matrix all the variances are of the same order except that of the third variate; evidently internode length is inherently more variable in these plants than are the other three characters measured.

Principal component analysis (PCA)

One purpose of many multivariate methods is to gain some idea of the relative importance of the different variates in a particular context. The technique known as Principal Component Analysis (PCA) is a simple multivariate method in terms of its theoretical background, but apart from its *relative* simplicity it can have some drawbacks, which are discussed later (pages 70–72).

The approach adopted here to explain the workings and interpretations of a principal component analysis will be to deal with the mathematics involved, including geometrical interpretations, using a bivariate situation. However, in practice the application of PCA to bivariate data is a trivial and pointless exercise; so to show how to interpret the results of a PCA applied to a biologically relevant situation we shall extend the mathematical concepts to four dimensions using the data of Example 2.1. First, however, we must consider the geometrical representation of the bivariate normal distribution.

The representation of a bivariate normal distribution by an ellipse

Representing the probability density curve of a univariate normal distribution requires two dimensions (Fig. 2.1a); to represent the **probability density surface** of a bivariate normal distribution, three dimensions are required (Fig. 2.1b). In both these diagrams the variate or variates (i.e. the quantity or quantities actually measured) comprise the horizontal axis or axes. In the bivariate case, therefore, the various combinations of values of the two variates can be found on the x_1-x_2 horizontal axis plane.

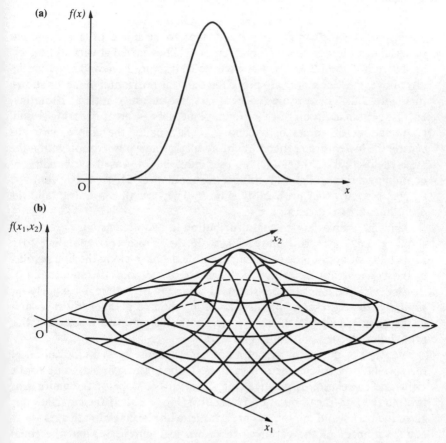

Figure 2.1 Normal distributions: (a) univariate; (b) bivariate.

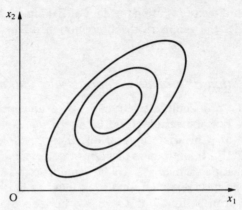

Figure 2.2 The representation of a bivariate normal distribution as concentric probability ellipses.

Imagine the probability density surface to be made of a transparent material and there to be horizontal 'contour' lines etched at various heights, as shown in Figure 2.1b. If now the whole structure is viewed from vertically above, these contours are projected onto the horizontal plane, as shown in Figure 2.2. These projected contours, which are elliptical in outline, enclose certain areas on the horizontal plane, that is, areas denoting certain combinations of values of x_1 and x_2. The smaller the ellipse (near the centre) the lower is the probability of finding a single observation within the area enclosed; but the larger ellipses have an increased probability of containing a single observation from the population. Hence these 'contour' ellipses are known as **probability ellipses.** They are all concentric, and the co-ordinates of the centre are (μ_1, μ_2).

Depicting a bivariate normal distribution in two dimensions necessitates showing only one probability ellipse. If the ellipse corresponding to a probability of about 0.68 is drawn, then the distance along the line parallel to the x_1-axis from the centre of the ellipse to the point of intersection of the line with the ellipse is the standard deviation of variate x_1, namely σ_1; similarly, the corresponding distance along the line parallel to the x_2-axis is the standard deviation of x_2, σ_2. So the advantage of selecting this probability ellipse is obvious (Fig. 2.3).

The slope of the major axis of a probability ellipse gives an indication of the correlation between the two variates. If the major axis slopes up to the right, the correlation is positive (Fig. 2.3a); if the slope of the major axis is up to the left, the correlation is negative (Fig. 2.3b). If the correlation is zero, the major and minor axes are parallel with the co-ordinate axes — unequal variances of the variates are shown in Figure 2.3c, and the equal variances case appears in Figure 2.3d.

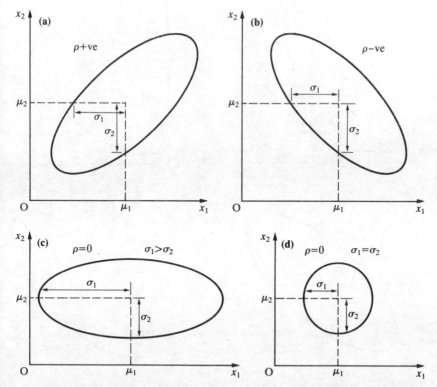

Figure 2.3 Elliptical representations of bivariate normally distributed populations: (a) with positive correlation between the variates; (b) with negative correlation; (c) with zero correlation and unequal variances; (d) with zero correlation and equal variances.

The mathematical basis of PCA

The fundamental principle of PCA is that the variates, which are usually correlated with one another, are changed into an equal number of uncorrelated components. The graphs in Figure 2.4 show the geometry of the processes involved in relation to a bivariate normally distributed population depicted by a single ellipse. In practice, however, a sample of observations is used, but if the sample is large enough, the vague outline of the swarm of points on a graph of the same will be seen to be elliptical (Fig. 2.4a). The population ellipse equivalent is shown in Figure 2.4b.

The first 'action' performed in PCA is the centralising of the population so that the centre of the ellipse lies at the origin. For the sample data, this is achieved by subtracting $\bar{x}_1$ from the x_1-value of each point and $\bar{x}_2$ from the x_2-value of each point (Fig. 2.4c). Then the axes are rotated about the

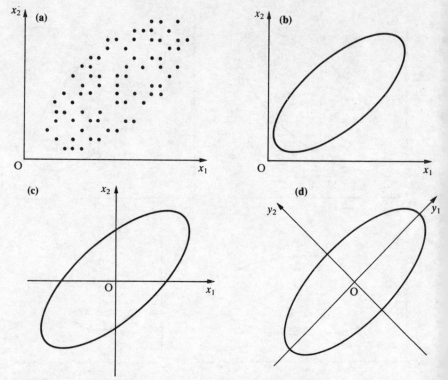

Figure 2.4 Representations of a bivariate normal distribution: (a) an artificial bivariate sample; (b) the corresponding bivariate population; (c) as (b), but centred on the origin; (d) as (c), but with axes rotated (components).

origin until they coincide with the major and minor axes of the ellipse (Fig. 2.4d). Because of this, the variates represented by the new axis positions (the components) are uncorrelated with one another.

Now consider a single point (observation), point P in Figure 2.5a. This point P is shown relative to the original variate axes, x_1 and x_2, and the new component axes, y_1 and y_2. Because P can be any point, the above symbols can also represent its co-ordinates with respect to each pair of axes. We now enquire into the relationship between the new co-ordinates, y_1 and y_2, and the old, x_1 and x_2.

In Figure 2.5a

$$y_1 = OM + MN + NQ$$

Now in triangle OMR, $OM = OR \cos \alpha = x_1 \cos \alpha$

In triangle MNR, $MN = NR \sin \alpha$

In triangle NPQ, $NQ = NP \sin \alpha$

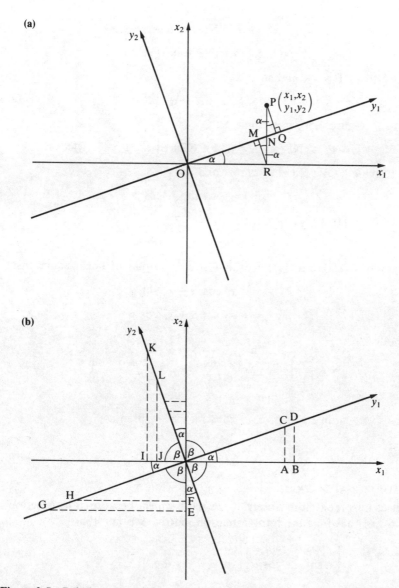

Figure 2.5 Relationships between quantities with respect to rotated axes: (a) relationships between the co-ordinates of a point with respect to original and rotated axes; (b) relationships between distances along the original axes and their projection onto rotated axes.

So

$$y_1 = x_1 \cos \alpha + NR \sin \alpha + NP \sin \alpha$$

$$= x_1 \cos \alpha + (NR + NP) \sin \alpha$$

But $NR + NP = x_2$, and so

$$y_1 = x_1 \cos \alpha + x_2 \sin \alpha \qquad (2.4)$$

Also in Figure 2.5a, in triangle NPQ

$$y_2 = PQ = NP \cos \alpha = (PR - NR)\cos \alpha = (x_2 - NR)\cos \alpha$$

In triangle NOR, $NR = x_1 \tan \alpha$, and so

$$y_2 = (x_2 - x_1 \tan \alpha)\cos \alpha$$

But $\tan \alpha = (\sin \alpha)/(\cos \alpha)$, so

$$y_2 = x_2 \cos \alpha - x_1 \sin \alpha \qquad (2.5)$$

Equations (2.4) and (2.5) can be written as a pair of linear equations:

$$y_1 = \quad x_1 \cos \alpha + x_2 \sin \alpha$$

$$y_2 = - x_1 \sin \alpha + x_2 \cos \alpha$$

Define

$$\mathbf{y} = \begin{bmatrix} y_1 \\ y_2 \end{bmatrix} \qquad \mathbf{x} = \begin{bmatrix} x_1 \\ x_2 \end{bmatrix} \qquad \mathbf{A} = \begin{bmatrix} \cos \alpha & \sin \alpha \\ -\sin \alpha & \cos \alpha \end{bmatrix}$$

and so the above pair of equations may be written as

$$\mathbf{y} = \mathbf{Ax} \qquad (2.6)$$

ORTHOGONAL MATRICES

Let us now look more closely at matrix $\mathbf{A}$ which, because of its function in (2.6), is known as the **transformation matrix**. We find that

$$\mathbf{AA}^T = \begin{bmatrix} \cos \alpha & \sin \alpha \\ -\sin \alpha & \cos \alpha \end{bmatrix} \begin{bmatrix} \cos \alpha & -\sin \alpha \\ \sin \alpha & \cos \alpha \end{bmatrix}$$

$$= \begin{bmatrix} \cos^2 \alpha + \sin^2 \alpha & -\cos \alpha \sin \alpha + \sin \alpha \cos \alpha \\ -\sin \alpha \cos \alpha + \cos \alpha \sin \alpha & \sin^2 \alpha + \cos^2 \alpha \end{bmatrix}$$

$$= \begin{bmatrix} 1 & 0 \\ 0 & 1 \end{bmatrix}$$

(see page 123 for the trigonometrical relationship in the leading diagonal).

Also it can be shown that $A^TA = I$. Hence for this matrix

$$A^T = A^{-1}$$

Further,

$$|A| = \cos^2 \alpha - (-\sin^2 \alpha) = \cos^2 \alpha + \sin^2 \alpha = 1$$

A matrix of this kind, in which $AA^T = A^TA = I$ and $|A| = 1$ is called an **orthogonal matrix**.

If we write the trigonometric elements of A in the usual element notation, we have

$$A = \begin{bmatrix} a_{11} & a_{12} \\ a_{21} & a_{22} \end{bmatrix}$$

and here $a_{22} = a_{11}$ and $a_{21} = -a_{12}$. Therefore

$$AA^T = \begin{bmatrix} a_{11} & a_{12} \\ -a_{12} & a_{11} \end{bmatrix} \begin{bmatrix} a_{11} & -a_{12} \\ a_{12} & a_{11} \end{bmatrix} = \begin{bmatrix} a_{11}^2 + a_{12}^2 & -a_{11}a_{12} + a_{12}a_{11} \\ -a_{12}a_{11} + a_{11}a_{12} & a_{12}^2 + a_{11}^2 \end{bmatrix}$$

$$= \begin{bmatrix} 1 & 0 \\ 0 & 1 \end{bmatrix}$$

which implies that

$$a_{11}^2 + a_{12}^2 = 1$$

and so also that

$$a_{21}^2 + a_{22}^2 = 1$$

Hence the sum of the squares of each row (and each column) is unity.

The reason for digressing into these properties of orthogonal matrix A, and replacing the trigonometric form of the elements by the general form, is that in problems involving more than two variates (and so a larger square matrix A) the trigonometric relationships do not strictly apply. However, the matrix A will still be orthogonal, and all the above properties extend to an orthogonal matrix of any order.

The variances of the new components

Without going into a rigorous proof we may now state the core of the mathematical basis of PCA. As already said, the prime 'action' performed is the conversion of variates (which are usually correlated) into an equal number of uncorrelated components by axis rotation. Because the new components are uncorrelated their variance–covariance matrix must be of the form

$$\Lambda = \begin{bmatrix} \lambda_1 & 0 \\ 0 & \lambda_2 \end{bmatrix} \tag{2.7}$$

where the λ_i are the variances of the new components and the covariance is, of course, zero. Furthermore, the PCA rotation has actually converted the variance–covariance matrix of the variates

$$\mathbf{S} = \begin{bmatrix} s_{11} & s_{12} \\ s_{21} & s_{22} \end{bmatrix}$$

to that of the components (eqn (2.7)); in other words, $\boldsymbol{\Lambda}$ is the diagonal form of $\mathbf{S}$. On page 37, (1.10) shows the method of doing this, and the equivalent equation in the present situation is

$$\mathbf{ASA}^T = \boldsymbol{\Lambda} \tag{2.8}$$

Thus, comparing (1.10) and (2.8) we see that the λ_i are the eigenvalues of matrix $\mathbf{S}$ and the *rows* of matrix $\mathbf{A}$ are its eigenvectors. So we have:

The eigenvalues of the variance–covariance matrix of the original variates are the variances of the new components, and the eigenvectors are the rows of the transformation matrix.

INTERPRETING THE NEW COMPONENTS

Now let us return to the transformation matrix $\mathbf{A}$ and study it in conjunction with Figure 2.5b. The angles β are equal to $(90 - \alpha)$, and it can be shown that $\cos(90 - \alpha) = \sin \alpha$; thus we have

$$\cos \beta = \sin \alpha$$

Matrix $\mathbf{A}$ can thus be rewritten as

$$\mathbf{A} = \begin{bmatrix} \cos \alpha & \cos \beta \\ -\cos \beta & \cos \alpha \end{bmatrix}$$

so all the elements are cosines of angles. Now, $\cos 0° = 1$ and $\cos 90° = 0$; all the angles have to lie between $0°$ and $90°$ and the smaller the angle the larger the cosine. Thus, for a particular set of data, the numerical values of the elements of $\mathbf{A}$ inform us about the angles between new components and old variates.

The rows of $\mathbf{A}$, the eigenvectors, relate to the new components, and the columns of $\mathbf{A}$ are identified with the old variates. Thus, for example, $-\cos \beta$ is associated with the relationship between new component 2 and old variate 1; β is, in fact, the angle between these two axes. Evidently the smaller the angle between a new component and an old variate the more closely does the component resemble the variate. Each of the new components are made up of a 'blend' of the two old variates, a linear combination in fact (see (2.4) and (2.5)), but the 'contribution' or **loading** of each of the latter to any one of the former will differ. When the angle between a component and a variate is small (when the cosine, and so the relevant element in **A**, is large) we say that there is a high loading of the variate on the component, and conversely for a large angle (= low loading).

The degree of resemblance between components and variates can be appreciated by a further study of Figure 2.5b. For the relationship between x_1 and y_1, consider two points on x_1 (A & B). By projecting at right angles to this axis to the y_1-axis we see that the distance CD on y_1, which 'corresponds' to distance AB on x_1, is almost the same. In the situation in Figure 2.5b the angle α is small, the loading (cosine) of x_1 on y_1 is high, and resemblance is close. Now consider the loading x_2 on y_1. The distance EF on the x_2-axis when projected onto the y_1-axis is distance GH which is much greater; the angle β is large, the loading (cosine) of x_2 on y_1 is low, and the resemblance is much less. The negative loading of x_1 on y_2 can also be appreciated from Figure 2.5b. The distance IJ involves an increase in magnitude of x_1 in moving from I to J, but projected on the y_2-axis the corresponding movement is from K to L which is a *decrease* in magnitude of y_2. All the other loadings in this situation are positive, implying that increase in magnitude of a variate, e.g. x_2, corresponds also to an *increase* in magnitude of a component, e.g. y_1. Thus, in summary, it can be seen that the loadings (i.e. the elements of the eigenvectors or, equivalently, the transformation matrix) can provide information concerning the nature of the new components.

Finally let us return to the eigenvalues – the variances of the new components. Figure 2.4c indicates that the variances of the variates are roughly similar to one another. However, the variances of the new components (Fig. 2.4d) are seen to be quite different: the variance of y_1 is much greater than that of y_2 (remember that the variance is the square of the standard deviation). This is a general situation in PCA: $\lambda_1 > \lambda_2 > \ldots \lambda_n$, where there are n variates and components. On the other hand,

$$s_{11} + s_{22} = \lambda_1 + \lambda_2 \qquad (2.9)$$

so the total variability of the system, as measured by the sum of the variances of the variates, is conserved; but it is repartitioned, so that $\lambda_1 > \lambda_2$ always.

A SIMPLE NUMERICAL EXAMPLE

We shall end this section by putting some numerical values to all the quantities we have been discussing, using the sample points shown in Figure 2.4a. The sample variance–covariance matrix is

$$\mathbf{S} = \begin{bmatrix} 18.5525 & 11.3278 \\ 11.3278 & 16.0186 \end{bmatrix}$$

So the variance of variate 1, s_{11}, is slightly greater than that of variate 2, s_{22}. After rotation, the variances of the new components (the eigenvalues of **S**) are

$$\lambda_1 = 28.6840 \text{ and } \lambda_2 = 5.8871$$

and we also find that

$$18.5525 + 16.0186 = 28.6840 + 5.8871 = 34.5711$$

the total variability of the system (eqn (2.9)). The transformation matrix is

$$\mathbf{A} = \begin{bmatrix} 0.7454 & 0.6667 \\ -0.6667 & 0.7454 \end{bmatrix}$$

which corresponds to angles $\alpha = 41.8°$ and $\beta = 48.2°$.

The two angles, and hence the loadings, are quite similar; and so each new component is made up of an almost equal contribution (in the absolute sense) of each variate. On the other hand, the variance of component 1 comprises 83% of the total variability. The significance of all this will become apparent in the next section in relation to real data.

The interpretation of PCA

The data set of Example 2.1 will be used. Since there are four variates, we require to work in four-dimensional space. First, however, let us consider briefly a three-variate problem, to bridge the transition from two to many dimensions. The three-dimensional analogue of the ellipse, which graphically describes a bivariate normal distribution, is an ellipsoid which depicts a trivariate normal distribution. An ellipsoid can be imagined as a flattened rugby football or as a *Paramecium* cell, and we further imagine the ellipsoid to be centred at the origin of a three-dimensional co-ordinate system; the three axes of the ellipsoid will be unlikely to coincide with any of the co-ordinate axes. Two independent rotations at most, in planes perpendicular to one another, will be required to bring the co-ordinate axes into correspondence with the ellipsoid axes; in the two-dimensional system there is only one independent rotation required. In general, an n-dimensional system accommodates $n - 1$ independent rotations. Once the rotations have been accomplished, the results and their interpretations are analagous to those developed above for the two-dimensional situation.

In the four-dimensional situation, implied by the data of Example 2.1, we can no longer graphically depict the system under study; but the mathematical theory of the two-dimensional case generalises without difficulty (in principle) to any number of dimensions, and so also does the mode of interpretation.

Example 2.2   Carry out a Principal Component Analysis on the data of Example 2.1 and interpret the results.

To perform a PCA means calculating the eigenvalues and eigenvectors of the variance–covariance matrix of the data. With four or more variates this is a job for a computer; so here the results will be merely presented, but the interpretation will require careful discussion.

The eigenvalues are presented first, in the table below, in descending order, together with the percentage contribution of each to the total variability and also the cumulative percentage of each.

	1	2	3	4
eigenvalue	0.2906	0.0855	0.0577	0.0151
percentage of total	64.30	19.58	12.77	3.34
cumulative percentage	64.30	83.89	96.66	100.0

Also we have

$$\sum_{i=1}^{4} \lambda_i = 0.4519 \quad \text{and} \quad \sum_{i=1}^{4} s_{ii} = 0.4518$$

The difference in the last place of decimals is simply due to rounding error in computing the eigenvalues. We see that component 1 accounts for some 64% of the total variability and that components 1 and 2 together account for almost 84%. In the context of this method, 'variation' is synonymous with 'information', and so we can say that the first two components together contain nearly 84% of the total information of the entire data set.

A major difficulty in trying to understand a body of multivariate data is the sheer bulk of it, particularly the fact that there *are* several variates which must be viewed and assimilated together. Any technique which can assist in this digestion is of value, and PCA does this in the following way. With four variates it is impossible to draw a graph of all the data involving all the variates simultaneously in order to get an idea of the data structure. One would have to draw graphs of the data two variables at a time; to show all the relationships among four variables would need six graphs. On the other hand, we have seen that the first two new components contain about 84% of the total information in the data set, and so a graph of these two will contain most of the available information and can be used on its own, if one is prepared to sacrifice the remaining 16% of the information contained in the data set. Thus it is possible for this set of data to draw a single graph which contains nearly 84% of the total information. The loss of relatively little information is a small price to pay for the advantage of essentially reducing the situation from four to two dimensions.

Not all data sets, however, can be realistically reduced to two dimensions. In general, the more variates the smaller is the proportion of the total variability accounted for by the first few components. Also, the percentage of the total variability that the investigator feels ought to be included is quite arbitrary; some will include enough components to cover at least 80%, whereas others will stop when they meet a component that by itself takes up less than about 5% of the total variability. Another consideration is whether a component can be interpreted in any way that makes biological sense. In the current example we will use the first two principal components.

We now want to interpret the principal components in terms of the old variates. For this we need the transformation matrix, **A**, in which the rows are the eigenvectors and contain the loadings of the old variates on the new components. The complete transformation matrix is

$$\mathbf{A} = \begin{bmatrix} 0.2434 & 0.3217 & 0.9136 & 0.0511 \\ 0.5382 & -0.3621 & -0.0584 & 0.7588 \\ -0.0771 & 0.8451 & -0.3014 & 0.4348 \\ 0.8032 & 0.2263 & -0.2667 & -0.4823 \end{bmatrix}$$

The first two eigenvectors relate to our principal components, and the four columns identify with the original variates for the purpose of examining loadings.

On the first component we see one outstandingly high loading – that of variate 3 (internode length). We conclude that component 1 is almost synonymous with internode length. Component 2, on the other hand, has no single dominant loading, but the positive loadings (leaf length and bract length) are numerically the highest. These two measurements are foliage lengths, so we may tentatively say that component 2 is one of foliage lengths but with an appreciable contribution of pedicel length in the opposite sense, that is, a contrast between foliage lengths and pedicel lengths. In the cases where there are many more variates it is useful to plot a graph of loadings, and this is done for the present example for the principal components in Figure 2.6a.

Our final aim is to plot the original observations on the principal component axes. To do this requires the implementation of equation (2.6). In this example the system for any one observation is

$$\begin{bmatrix} y_1 \\ y_2 \\ y_3 \\ y_4 \end{bmatrix} = \begin{bmatrix} 0.2434 & 0.3217 & 0.9136 & 0.0511 \\ 0.5382 & -0.3621 & -0.0584 & 0.7588 \\ -0.0771 & 0.8451 & -0.3014 & 0.4348 \\ 0.8032 & 0.2263 & -0.2667 & -0.4823 \end{bmatrix} \begin{bmatrix} x_1 \\ x_2 \\ x_3 \\ x_4 \end{bmatrix}$$

but the x_is are not directly the measurements of the observation in hand.

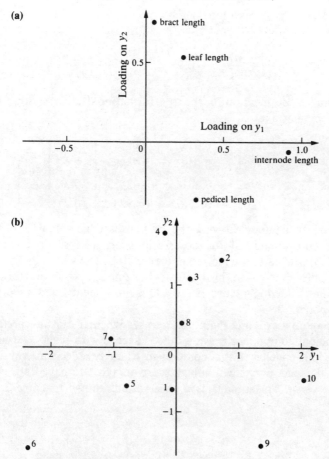

Figure 2.6 Results of the Principal Component Analysis of Examples 2.1 and 2.2: (a) variate loadings; (b) observation scores.

Instead, they are *standardised* measurements obtained by subtracting the appropriate mean from each measurement and dividing by its standard deviation. It is convenient to state the column vector of means, $\bar{x}$, and a column vector of standard deviations, s, the elements of the latter being the square root of each element in the leading diagonal of the variance–covariance matrix.

$$\bar{x} = \begin{bmatrix} 2.79 \\ 0.61 \\ 2.62 \\ 1.92 \end{bmatrix} \qquad s = \begin{bmatrix} 0.2300 \\ 0.2893 \\ 0.4991 \\ 0.2571 \end{bmatrix}$$

Taking the first observation in the data table on page 53, we have for example

$$x_1 = \frac{2.57 - 2.79}{0.2300} = -0.96 \qquad x_3 = \frac{2.77 - 2.62}{0.4991} = 0.30$$

The complete specification for the transformation of observation 1 onto the new component axes is

$$\begin{bmatrix} 0.2434 & 0.3217 & 0.9136 & 0.0511 \\ 0.5382 & -0.3621 & -0.0584 & 0.7588 \\ -0.0771 & 0.8451 & -0.3014 & 0.4348 \\ 0.8032 & 0.2263 & -0.2667 & -0.4823 \end{bmatrix} \begin{bmatrix} -0.96 \\ -0.28 \\ 0.30 \\ -0.31 \end{bmatrix} = \begin{bmatrix} -0.05 \\ -0.67 \\ -0.38 \\ -0.75 \end{bmatrix}$$

Below are two tables showing the standardised data matrix (Table 2.2) and the co-ordinates of the observations on the principal components (Table 2.3); the latter are plotted in Figure 2.6b. The values in Table 2.3 are usually called 'scores' in this context; thus one speaks about, for example, observation 4 having a score of -0.19 on component 1 and a score of 1.82 on component 2.

We may now examine Figure 2.6b to see whether our interpretations of the principal component axes do indeed accord with reality. According to our previous conclusion that component 1 largely reflects internode length, we would expect plants with a high score on axis 1 to have long internodes and vice versa. The standardised observations shown in Table 2.2 will be

Table 2.2 The data matrix of Table 2.1 in standardised form.

Observation	Variate			
	1	2	3	4
1	-0.96	-0.28	0.30	-0.31
2	0.91	0.10	0.42	1.28
3	0.65	-0.48	0.18	0.78
4	1.09	0.10	-0.64	1.63
5	-0.09	-1.49	-0.26	-1.44
6	-2.13	0.10	-2.02	-0.70
7	0.17	-0.93	-0.84	-0.43
8	-0.35	-0.48	0.30	0.58
9	-0.55	1.90	0.94	-0.78
10	0.91	1.31	1.56	-0.55

Table 2.3 The scores of each observation of the data in Table 2.2 on the two principal components.

Observation	Principal component 1	2
1	−0.05	−0.67
2	0.72	1.40
3	0.22	1.10
4	−0.19	1.82
5	−0.80	−0.59
6	−2.36	−1.60
7	−1.04	0.15
8	0.08	0.41
9	1.36	−1.53
10	2.04	−0.49

useful here, since we can see at a glance those values which exceed their mean (positive) and those that are less than their mean (negative).

On component 1, plant 10 has the highest score, and we also see from Table 2.2 that it has the longest internode. Plant 9 has the second highest score and indeed has the second longest internode measurement. At the other end of the scale, plant 6 has the lowest score on component 1 and is also seen to have the shortest internode length; plant 7 has the second lowest score and also the second shortest internode measurement. For the extreme-scoring plants on axis 1, therefore, there is a one-to-one correspondence with relative internode lengths; this is expected because the loading of inter-node length on component 1 is so outstandingly high.

On component 2, in which leaf and bract lengths are moderately highly loaded, and contrast with pedicel length, plant 4 has the highest score, and it also has the longest leaf and bract lengths but only a medium pedicel length; plant 2 has the second highest score, and it has the second highest leaf length and the second highest bract length, again combined with a medium pedicel length. At the other end, plant 6 has the lowest score on component 2, and this plant also has the smallest leaf measurement but the second smallest bract length, and, once more, a medium pedicel length; plant 9 has the second smallest score, the third smallest leaf length, and the second smallest bract length but the highest pedicel length. There is not quite the same degree of relative correspondence between component 2 scores and foliage lengths as there is between component 1 scores and inter-node lengths, but the loadings on the second component are not as outstandingly clear cut as is the loading of internode length on the first component. Also, because of the lower absolute value of the loading of pedicel length on component 2, the contrast between pedicel and foliage lengths is rather weak.

Apart from any particular interpretations of the principal components that may be made in relation to a specific set of data, one rather more general inference can often be made. If all the original variates are size measurements, and if all the loadings on the first principal component have the same sign, then the first principal component is a general size measure. Thus individual observations which have high scores on principal component 1 tend to be of large size if the loadings are positive (small size if the loadings are negative) and vice versa. This is indeed the case for our *Silene maritima* example, since the elements of the first row of the transformation matrix, **A**, are all positive (page 66). If Table 2.3 is compared with Table 2.2 it is evident that plant 10, which has the highest score on principal component 1, generally has large measurements; plant 6, which has the lowest score on component 6, has mostly small measurements, and similarly for the other plants.

In conclusion, the achievement of PCA in relation to these data is that the latter have been effectively summarised in Figure 2.6b with the loss of only about 16% of the total information. The meanings of the principal components are indicated in Figure 2.6a. Besides summarising the leading features of the data set, PCA has also indicated which are the characters of importance in this situation. For an initial survey, such a pointer can be useful in saving unnecessary work later on. Notice that the word 'indication' has been used more than once in the above; also you may have noticed that unlike the situation in univariate statistical methods where a precise statement can be made at a specified level of probability, in multivariate situations it is not usually possible to make such definite statements. Multivariate methods indicate features and trends in the data, but the final interpretation is often somewhat subjective.

Problems in using PCA

In Examples 2.1 and 2.2 the variates are all of the same type − linear measurements − and expressed in the same units − mm. Nevertheless, a problem still exists in relation to using PCA, namely, that the organs measured differ markedly in size, from the pedicels (smallest) to leaves and internodes (largest). Now it is an almost universal phenomenon in biology that large measurements tend to have large variances and vice versa, and the present data are no exception.

Although PCA does not require the data to approximate to a multivariate normal distribution, the method does work better if this is the case. Further, as for most other statistical methods, the different variates should be of constant variance. Variates whose variances (or more strictly, standard deviations) are roughly proportional to their means approximate to a lognormal distribution. This implies that the logarithms of the variate values approximate to a normal distribution. Not only does the logarithmic transformation of such multivariate data tend to make them approximate to a multivariate normal distribution, it also tends to equalise the variances. A logarithmic transformation of the data of the present example was made prior to their presentation in Example 2.1. Even so, a glance at the variance–covariance matrix of the transformed data, S (page 54), shows that the variance of internode length is markedly greater than the others, although the difference is not statistically significant. Further, we found in the transformation matrix, A (page 66), that internode length dominates the first eigenvector. This is a general feature of PCA: variates with high variances tend to dominate the analysis.

Now, without logarithmic transformation the order of magnitude of the variances is almost the same as the order of magnitude of the means. A PCA carried out on the untransformed data yields eigenvectors which reflect this: the first eigenvector is dominated by internode length (largest variance), the second eigenvector is dominated by leaf length (second largest variance), down to the last eigenvector which is dominated by pedicel length which has the smallest variance. Clearly, nothing has been gained by this analysis.

Logarithmic transformation of the data still leaves one outstandingly high variance – that of internode length; but when it is realised that this does not obviously reflect the fact that internode lengths are relatively large, because leaf lengths are of the same order of magnitude, then one can only conclude that internode length is inherently more variable than the other three characters measured on these plants. In the PCA based on the logarithmically transformed data, discussed in Example 2.2, internode length does indeed dominate the first component, but this obviously has a more deep-seated meaning than the same situation arising in a PCA based on untransformed data.

In practice, many situations arise in which a body of multivariate data consists of variates of different types: some may be lengths, some weights, others may even be physiological measurements. Under these circumstances variance heterogeneity, seriously affecting the PCA, is almost bound to arise. The commonest way of evading this difficulty is based on the following reasoning.

From a mathematical viewpoint, the central equation in PCA is (2.8), showing the relationship between the eigenvalues, the eigenvectors, and the variance–covariance matrix. Equation (2.8) further shows that both the matrix of eigenvalues and the matrix of eigenvectors are directly derived from the variance–covariance matrix (cf. page 37); hence we usually say that the PCA is done on the variance–covariance matrix. Now, the correlation matrix is identical in structure to the variance–covariance matrix and may replace the variance–covariance matrix in the analysis. We then speak of doing a PCA on the correlation matrix. How does this help our current problem, namely, variance heterogeneity?

An illuminating way of interpreting the correlation matrix is that it is the variance–covariance matrix of **standardised variates.** (To transform an observation from a given variate into the equivalent observation from the standardised variate, one merely divides the former by the standard deviation of that variate.) Since all the 'variances' are now the same – unity – it implies that all variates are of equal importance if the correlation matrix is submitted to a PCA.

The question of which matrix to submit to PCA largely depends upon the nature of the data in hand, and the inferences to be drawn from the data. Where the variates are of a similar kind, with variances which do not show obvious differences (e.g. change with overall size) as in Example 2.1, then the variance–covariance matrix should be used; indeed, PCA performed on the correlation matrix of the data of Example 2.1 was not as revealing as the analysis carried out on the variance–covariance matrix in Example 2.2. On the other hand, since most multivariate data are not as 'homogeneous' as the above example, it is not surprising to find that most PCAs are carried out using the correlation matrix. Such a procedure is not without problems of its own and these are well discussed by Seal (1964, pp. 117–20).

Other multivariate methods

The mathematics of principal component analysis is similar to the theoretical basis of several other multivariate methods. As with univariate procedures, an appropriate multivariate method is chosen on the basis of data structure and the questions to be asked of the data: principal component analysis is only one of several multivariate statistical methods, but many others are based on concepts and theory described above in relation to PCA. An excellent introduction is given by Chatfield and Collins (1980).

The stable age distribution in population growth

This section follows on from the part of Chapter 11 in Causton (1983) entitled 'Population growth with age-dependent birth and death rates', to which you should refer. Equation (11.21) there reads

$$n_{t+s} = P^s n_t \qquad (2.10)$$

This equation means that, starting with an initial population at time t which is represented by a column vector, n_t, whose elements give the number of individuals in each age class, then after s time intervals the structure of the population is given by the analogous column vector $\mathbf{n}_{t+s}$. Matrix $\mathbf{P}$ contains information on probabilities of survival of the different age classes and the average number of offspring produced by individuals of different ages and is known as the transition matrix. Evidently, for several time periods, s, we are raising the transition matrix to the power s.

Now whatever the actual values of the elements of $\mathbf{P}$ and $\mathbf{n}_t$, if the number of time intervals in which equation (2.10) applies becomes large, then the ratios of the elements in $\mathbf{n}_{t+s}$ tend towards steady values. In other words, as s increases, the proportional age distribution of the population tends to become stable. Thus for large s we have

$$\mathbf{n}_{t+s+1} \propto \mathbf{n}_{t+s} \tag{2.11}$$

and so, using (2.10), we have

$$\mathbf{Pn}_{t+s} \propto \mathbf{n}_{t+s}$$

Replacing the proportionality sign by a constant gives

$$\mathbf{Pn}_{t+s} = \lambda\mathbf{n}_{t+s} \tag{2.12}$$

Equation (2.12) shows that to find the stable age distribution of a population which grows according to the model (2.10) simply requires finding eigenvalues and eigenvectors of the transition matrix $\mathbf{P}$. The constant of proportionality, λ, is an eigenvalue, and $\mathbf{n}_{t+s}$ is the corresponding eigenvector whose elements are the stable age distribution (i.e. the elements show what *proportion* of the population falls into each age category).

The above is the theory. In practical terms one must ask *which* of the several eigenvalues and eigenvectors are relevant to our problem.

We first note that the transition matrix $\mathbf{P}$ is an example of a **positive–regular matrix.** This implies that it is non-negative (all its elements are $\geqslant 0$), non-singular, and irreducible. This last property means that it is impossible to interchange rows with rows and columns with columns to produce another matrix which can be partitioned into square submatrices of order > 1. Possession of these properties means that $\mathbf{P}$ meets the necessary conditions for the Perron–Frobenius theorem to be applicable. As a result of this theorem we know that $\mathbf{P}$ will have at least one real positive eigenvalue; it is the largest such eigenvalue in which we are interested (indeed, it may be the only real positive eigenvalue) and is known as the maximum or Perron latent root. The eigenvector corresponding to this eigenvalue has elements all of the same sign (which may be taken as positive) and is the only such eigenvector. This eigenvector, of course, has the properties we require: all the proportions of the population at the different ages must be positive.

The final point of interest is that the Perron root is the rate of natural increase of the population when it has reached the stable age distribution, or the absolute growth rate, as is obvious from equation (2.12); hence the requirement for this eigenvalue to be real and positive.

Example 2.3   In this example the numbers have been deliberately chosen to keep the arithmetic simple. Calculate the stable age distribution of a population whose transition matrix is given by

$$\mathbf{P} = \begin{bmatrix} 1 & 3 & 4 & 12 \\ \frac{1}{2} & 0 & 0 & 0 \\ 0 & \frac{1}{4} & 0 & 0 \\ 0 & 0 & \frac{2}{3} & 0 \end{bmatrix}$$

Calculate also the absolute growth rate of the population at the time.

In effect, we must answer the question in reverse. Since the growth rate we require is the Perron eigenvalue, we must first solve the characteristic equation, which is

$$|\mathbf{P} - \lambda\mathbf{I}| = \begin{vmatrix} 1-\lambda & 3 & 4 & 12 \\ \frac{1}{2} & -\lambda & 0 & 0 \\ 0 & \frac{1}{4} & -\lambda & 0 \\ 0 & 0 & \frac{2}{3} & -\lambda \end{vmatrix} = 0 \qquad (2.13)$$

Expanding the determinant gives

$$(1-\lambda)\begin{vmatrix} -\lambda & 0 & 0 \\ \frac{1}{4} & -\lambda & 0 \\ 0 & \frac{2}{3} & -\lambda \end{vmatrix} - \frac{1}{2}\begin{vmatrix} 3 & 4 & 12 \\ \frac{1}{4} & -\lambda & 0 \\ 0 & \frac{2}{3} & -\lambda \end{vmatrix} = 0$$

(see page 6).

Multiplying out gives the following quartic equation in λ:

$$\lambda^4 - \lambda^3 - \tfrac{3}{2}\lambda^2 - \tfrac{1}{2}\lambda - 1 = 0$$

Solving a quartic equation is very difficult. To start with, one should sketch the curve of the equation $y = \lambda^4 - \lambda^3 \frac{3}{2}\lambda^2 - \frac{1}{2}\lambda - 1$ by plotting for some integer values of λ greater and less than zero (see page 183). By this means, some idea will be gained concerning the ranges of λ-values in which the roots occur.

In this case, however, we should immediately see that $\lambda = 2$ is a root of this equation by sketching the curve; hence by taking out -2 as a factor of the left-hand side we get

$$(\lambda - 2)(\lambda^3 + \lambda^2 + \tfrac{1}{2}\lambda + \tfrac{1}{2}) = 0$$

Inspection of the cubic bracket, again by sketching the curve, shows that $\lambda = -1$ is a root and so factorising again gives

$$(\lambda - 2)(\lambda + 1)(\lambda^2 + \tfrac{1}{2}) = 0$$

The third bracket implies that $\lambda^2 = -\tfrac{1}{2}$ or

$$\lambda^2 = \tfrac{1}{2}(-1)$$

Now put $i^2 = -1$ (see page 163), then

$$\lambda^2 = \tfrac{1}{2}i^2$$

which finally yields

$$\lambda = \pm \frac{1}{\sqrt{2}} i$$

So the four eigenvalues are

$$2, \ -1, \ \pm \frac{1}{\sqrt{2}} i.$$

The first eigenvalue is the Perron root and is the only eigenvalue which is both real and positive. Hence the absolute growth rate at stable age distribution is two individuals per unit of time.

To find the corresponding eigenvector we use the matrix equation

$$[\mathbf{P} - \lambda\mathbf{I}]\mathbf{x} = \mathbf{0} \tag{2.14}$$

with $\lambda = 2$. We can find the expanded version of (2.14) by substituting for λ in (2.13), giving

$$\begin{bmatrix} -1 & 3 & 4 & 12 \\ \tfrac{1}{2} & -2 & 0 & 0 \\ 0 & \tfrac{1}{4} & -2 & 0 \\ 0 & 0 & \tfrac{2}{3} & -2 \end{bmatrix} \begin{bmatrix} x_1 \\ x_2 \\ x_3 \\ x_4 \end{bmatrix} = \begin{bmatrix} 0 \\ 0 \\ 0 \\ 0 \end{bmatrix}$$

and writing out in equation form yields

$$-x_1 + 3x_2 + 4x_3 + 12x_4 = 0 \tag{1}$$

$$\tfrac{1}{2}x_1 - 2x_2 \qquad\qquad = 0 \tag{2}$$

$$\tfrac{1}{4}x_2 - 2x_3 \qquad = 0 \tag{3}$$

$$\tfrac{2}{3}x_3 - 2x_4 = 0 \tag{4}$$

These equations are immediately solvable. Put $x_4 = \mu$, then from (4)

$x_3 = 3\mu$, from *(3)* $x_2 = 24\ \mu$, and from *(2)* $x_1 = 96\ \mu$. Checking in *(1)* that the system is consistent gives $-96\ \mu + 72\ \mu + 12\ \mu + 12\ \mu = 0$. So the solution is given by

$$\begin{bmatrix} x_1 \\ x_2 \\ x_3 \\ x_4 \end{bmatrix} = \begin{bmatrix} 96 \\ 24 \\ 3 \\ 1 \end{bmatrix}$$

and the elements in the vector are the proportions of each age group in the population (out of a total of $96 + 24 + 3 + 1 = 124$). The actual proportions are $\frac{24}{31}, \frac{6}{31}, \frac{3}{124}, \frac{1}{124}$, or 0.7742, 0.1936, 0.0242, 0.0081.

A much fuller account of population growth using matrices is given in Pielou (1977, Ch. 3).

Biological applications of linear equations

Applications of linear equations in biology are diverse. Below are given two kinds of example; both are based on examples from Grossman and Turner (1974).

Example 2.4   The activities of a grazing animal can be classified roughly into three categories: grazing, moving (to new grazing areas or to avoid predators), and resting. The net energy gain (above maintenance requirements) from grazing is 200 cal h^{-1}. The net energy losses in moving and resting are 150 and 50 cal h^{-1}, respectively.

(a)   *How should the day be divided among the three activities so that the energy gains during grazing exactly compensate for energy losses during moving and resting?*
(b)   *Is this division of the day unique?*
(c)   *If the animal must rest for at least 6 hours every day, how should the day be divided?*
(d)   *If, to avoid overgrazing, the animal must spend equal times moving and grazing, how should the day be divided?*

(a & b)   Denote the time in hours utilised for grazing as g, for resting as r, and for moving as m. Because the total of g, r and m must equal 24 hours, we can write $g + r + m = 24$. For the energy component of the problem we have $200\ g - 50r - 150\ m = 0$. So we have a pair of linear equations and no other information is available.

$$g + r + m = 24$$
$$200\ g - 50\ r - 150\ m = 0$$
(2.15)

Hence part (b) is answered; there can be no unique relationship, and solution of two of the variables depends on setting values of the third. Movement is the least predictable of the three variables, in the absence of any further information; it depends on the amount of predation and the distance necessary to travel in order to obtain adequate grass. Solving in the usual way and putting $m = \lambda$ gives

$$\begin{bmatrix} g \\ r \\ m \end{bmatrix} = \lambda \begin{bmatrix} 0.4 \\ -1.4 \\ 1.0 \end{bmatrix} + \begin{bmatrix} 4.8 \\ 19.2 \\ 0 \end{bmatrix}. \qquad (2.16)$$

Now, in theory, λ can take any value, but if it is too high then r will become negative, which is clearly inadmissible in this context. From the vector equation (2.16) above, we have

$$g = 4.8 + 0.4\ \lambda \qquad (2.17)$$

and

$$r = 19.2 - 1.4\ \lambda \qquad (2.18)$$

Put $r = 0$, i.e. assume that there is no resting period; then from (2.18) $\lambda \simeq 13.71$. Substituting into (2.17) gives $g = 10.28$; so $g + \lambda = 23.99$, i.e. 24 hours aside from rounding error. So the maximum the moving period can be is 13.71 hours, and so we write, in addition to (2.16), $0 \leqslant \lambda \leqslant 13.71$ for the solution of (2.15).

(c)   Put $r = 6$ in (2.18), then $\lambda \simeq 9.43$. Substitution in (2.17) then gives $g = 8.57$. So if the rest period is just 6 hours, then grazing is 8.57 hours and moving is 9.43 hours.

(d)   Put $g = \lambda$ in (2.17), then $\lambda = g = 8$ hours; so therefore all three periods must be equal at 8 hours each.

---

*Example 2.5   Three species of bacteria coexist in a test tube, and three kinds of nutrient are supplied. Define matrix **A** where element a_{ij} denotes the average consumption per day of the jth nutrient by an individual of the ith species. In the present instance*

$$\mathbf{A} = \begin{bmatrix} 1 & 1 & 1 \\ 1 & 2 & 3 \\ 1 & 3 & 5 \end{bmatrix}$$

If 20 000 units of nutrient 1, 30 000 units of nutrient 2, and 40 000 units of nutrient 3 are supplied daily, and all resources are consumed, what populations of the three species can coexist in this environment? Are these population unique?

Let x_i be the population of the ith species, then we have

$$x_1 + x_2 + x_3 = 20\ 000$$
$$x_1 + 2x_2 + 3x_3 = 30\ 000$$
$$x_1 + 3x_2 + 5x_3 = 40\ 000$$

Anticipating somewhat, we can find that $|A| = 0$, and so the solutions are not unique. By solving the above equations in the usual way and letting $x_3 = \lambda$, we find

$$\begin{bmatrix} x_1 \\ x_2 \\ x_3 \end{bmatrix} = \lambda \begin{bmatrix} 1 \\ -2 \\ 1 \end{bmatrix} + \begin{bmatrix} 10\ 000 \\ 10\ 000 \\ 0 \end{bmatrix}$$

From the above vector solution we have in particular that $x_2 = 10\ 000 - 2\lambda$. Now if λ is sufficiently large, x_2 would be negative. In practice, $x_2 \not< 0$, so put $x_2 = 0$ in the above equation, yield $\lambda = 5000$. So in the vector solution we have the constraint that $0 \leqslant \lambda \leqslant 5000$.

Suggestions for further reading, and other references

A good modern book on multivariate methods, with an excellent balance of theory and applications together with a review of other books in the field, is **Chatfield, C. and A. J. Collins** (1980). *Introduction to multivariate analysis*. London: Chapman & Hall.

General biological applications of multivariate methods, much more diverse than the title suggests, are given in **Blackith, R. E.** and **R. A. Reyment** (1971). *Multivariate morphometrics*. London: Academic Press.

Applications of multivariate methods in ecology are given by
Causton, D. R. (1987). *An introduction to vegetation analysis: principles, practice, and interpretation*. London: Allen & Unwin.
Gaugh, H. G. Jr (1982). *Multivariate analysis in community ecology*. Cambridge: Cambridge University Press.
Pielou, E. C. (1977). *Mathematical ecology*. New York: Wiley.

Both multivariate theory and computational details, together with extensive computer programs and a good chapter on matrix algebra, are given in **Mather, F. M.** (1976). *Computational methods of multivariate analysis in physical geography*. New York: Wiley.

Examples of linear equations in biology appear in **Grossman, S. I.** and **J. E. Turner** (1974). *Mathematics for the biological sciences, Ch.3*. London: Macmillan.

Other references:

Gower, J. C. (1966). Some distance properties of latent root and vector methods used in multivariate analysis. *Biometrika* **53**, 325–38.

Gower, J. C. (1967). Multivariate analysis and multidimensional geometry. *The Statistician* **17**, 13–28.

Leslie, P. H. (1945). The use of matrices in certain population mathematics. *Biometrika* **33**, 183–212.

Leslie, P. H. (1948). Some further notes on the use of matrices in population mathematics. *Biometrika* **35**, 213–45.

Seal, H. L. (1964). *Multivariate statistical analysis for biologists.* London: Methuen.

3

Integrals

Much of this chapter will be concerned with various kinds of integral and their application; but first two commonly employed methods of integration will be presented. One method – integration by partial fractions – has been dealt with in Causton (1983); here, the methods of integration by substitution and integration by parts will be described.

Two methods of integration

Integration by substitution

Suppose we were required to find $\int x\sqrt{(1 + x^2)} \, dx$. It is not possible to multiply out the terms first in order to look for a standard integral. However, we try to put part of the integrand equal to another variable in the hope that the resulting integral will be a standard one. In this example put

$$z = 1 + x^2 \tag{3.1}$$

Now, apart from the integral sign, the integral in this example consists of three terms: x and $\sqrt{(1 + x^2)}$, which together comprise the integrand, and the variable of integration, dx. We cannot merely substitute for one of these components, e.g. $\sqrt{(1 + x^2)}(=\sqrt{z})$, ending up with an integral containing the two variable quantities, x and z; but by substituting for each of the terms in the integral we must finally obtain an integral entirely in terms of z. Substituting for $\sqrt{(1 + x^2)}$ is easy: it is merely $\sqrt{z}$. Now from (3.1), $x = \pm\sqrt{(z-1)}$. Differentiating (3.1), $dz/dx = 2x$, i.e. $dx = dz/2x$. But we already have seen that $x = \pm\sqrt{(z-1)}$; hence $dx = \pm dz/2\sqrt{(z-1)}$. Thus we may write

$$\int x\sqrt{(1 + x^2)} \, dx = \int \sqrt{(z-1)}\sqrt{z} \, \frac{dz}{2\sqrt{(z-1)}} = \int \frac{\sqrt{z} \, dz}{2}$$

$$= \frac{1}{2} \int z^{1/2} \, dz$$

The last integral is a standard one and gives

$$\frac{1}{2}\left(\frac{z^{3/2}}{3/2}\right) + c = \frac{z^{3/2}}{3} + c = \frac{\sqrt{z^3}}{3} + c.$$

Substituting back to x for z, we have

$$\int x\sqrt{(1+x^2)}\, dx = \tfrac{1}{3}\sqrt{(1+x^2)^3} + c$$

Integration by substitution is not always a successful technique in practice; it is often difficult to know exactly what to substitute for what, and after many attempts at different substitutions it may be found that the integral cannot be reduced to a standard form by this method.

Example 3.1   *Find*

$$\int \frac{(x+1)\, dx}{\sqrt{(x^2+2x-1)}}$$

Put $z = (x^2+2x-1)^{1/2}$. By the function of a function rule we have

$$\frac{dz}{dx} = \frac{(x+1)}{(x^2+2x-1)^{1/2}}$$

i.e.

$$(x+1)\, dx = (x^2+2x-1)^{1/2}\, dz$$

Hence

$$\int \frac{(x+1)\, dx}{(x^2+2x-1)^{1/2}} = \int \frac{(x^2+2x-1)^{1/2}\, dz}{(x^2+2x-1)^{1/2}} = \int dz = z + c$$

and so, substituting back for z, we have the result

$$\int \frac{(x+1)\, dx}{\sqrt{(z^2+2x-1)}} = \sqrt{(x^2+2x-1)} + c$$

Integration by parts

This method is used when it is required to integrate the product of two functions of x, but the procedure can also be used to integrate certain single functions of x for which no other method suffices. The method involves a formula into which the function, or functions, are substituted in a similar manner to the product rule for differentiation. Indeed, we start with the product rule for differentiation to derive the formula for integration by parts.

81

For two functions of x, $\psi(x)$ and $\theta(x)$, the product rule for differentiation gives

$$\frac{d\{\psi(x)\theta(x)\}}{dx} = \psi(x)\frac{d\{\theta(x)\}}{dx} + \theta(x)\frac{d\{\psi(x)\}}{dx}$$

(cf. Causton 1983, eqns (5.20) on page 77).

Rearranging:

$$\psi(x)\frac{d\{\theta(x)\}}{dx} = \frac{d\{\psi(x)\theta(x)\}}{dx} - \theta(x)\frac{d\{\psi(x)\}}{dx}$$

Integrating with respect to x gives

$$\int \psi(x)\frac{d\{\theta(x)\}}{dx}\,dx = \psi(x)\,\theta(x) - \int \theta(x)\frac{d\{\psi(x)\}}{dx}\,dx + c$$

Put

$$\phi(x) = \frac{d\{\theta(x)\}}{dx}, \text{ then } \theta(x) = \int \phi(x)\,dx$$

The previous line now finally becomes

$$\int \psi(x)\phi(x)\,dx = \psi(x)\int \phi(x)\,dx - \int\left[\frac{d\{\psi(x)\}}{dx}\int \phi(x)\,dx\right]dx + c$$

$$(3.2)$$

Equation (3.2) is the formula for integration by parts. On examining the terms on the right-hand side of (3.2), we see that $\int\phi(x)\,dx$ appears in each term; consequently, $\phi(x)$ must be integrable by some other method. The function $\psi(x)$ is not required to be integrated: true, it is necessary to differentiate it, but as all functions can be differentiated this is no problem. Hence $\psi(x)$ may be a function which cannot be integrated by other methods.

In using the product rule to differentiate the product of two functions of x, it does not matter which function is identified with $\psi(x)$ or with $\phi(x)$ (see Causton 1983, eqn 5.20, p. 77); but when integrating by parts, the remarks in the previous paragraph must be borne in mind.

Example 3.2   Find

(a) $\int x\sqrt{(2x+3)^3}\,dx$

(b) $\int (x^2+1)\sqrt{(2x+3)}\,dx$

(a) Put $\psi(x) = x$ and $\phi(x) = \sqrt{(2x+3)^3}$; then using 3.2:

$$\int x\sqrt{(2x+3)^3}\,dx = x\int \sqrt{(2x+3)^3}\,dx - \int \left\{ \frac{d(x)}{dx} \int \sqrt{(2x+3)^3}\,dx \right\} dx$$

$$(3.3)$$

Now

$$\int \sqrt{(2x+3)^3}\,dx = \int (2x+3)^{5/2}\,dx$$

$$= \frac{(2x+3)^{7/2}}{2(5/2)} + c$$

$$= \frac{(2x+3)^{7/2}}{5} + c$$

and $d(x)/dx = 1$. Hence (3.3) becomes

$$\int x\sqrt{(2x+3)^3}\,dx = \frac{x(2x+3)^{7/2}}{5} - \int \frac{(1)(2x+3)^{7/2}\,dx}{5} + c$$

$$= \frac{x(2x+3)^{7/2}}{5} - \frac{(2x+3)^{9/2}}{35} + c$$

Thus

$$\int x\sqrt{(2x+3)^3}\,dx = \frac{x\sqrt{(2x+3)^5}}{5} - \frac{\sqrt{(2x+3)^7}}{35} + c \qquad (3.4)$$

(b) Put $\psi(x) = x^2 + 1$ and $\phi(x) = \sqrt{(2x+3)}$; then using (3.2):

$$\int (x^2+1)\sqrt{(2x+3)}\,dx = (x^2+1)\int \sqrt{(2x+3)}\,dx$$

$$- \int \left\{ \frac{d(x^2+1)}{dx} \int \sqrt{(2x+3)}\,dx \right\} dx \quad (3.5)$$

Now

$$\int \sqrt{(2x+3)}\,dx = \int (2x+3)^{1/2}\,dx = \frac{(2x+3)^{3/2}}{3} + c$$

and

$$\frac{d(x^2+1)}{dx} = 2x$$

Hence (3.5) becomes

$$\int (x^2+1)\sqrt{(2x+3)}\,dx = \frac{(x^2+1)(2x+3)^{3/2}}{3} - \int \frac{2x(2x+3)^{3/2}}{3} + c$$

i.e.

$$\int (x^2+1)\sqrt{(2x+3)}\ dx = \frac{(x^2+1)(2x+3)^{\frac{3}{2}}}{3} - \tfrac{2}{3}\int x(2x+3)^{\frac{3}{2}} + c$$
(3.6)

This example is of a type that frequently arises in integration by parts, i.e. the process has to be done twice, or even more. Here the integral in (3.6) has to be integrated by parts again; however, in this instance it has already been done in (a), (3.4), so we have the result

$$\tfrac{2}{3}\int x(2x+3)^{\frac{3}{2}}\ dx = \frac{2x(2x+3)^{\frac{5}{2}}}{15} - \frac{2(2x+3)^{\frac{7}{2}}}{105} + c$$

and so, substituting in (3.6), we finally have

$$\int (x^2+1)\sqrt{(2x+3)}\ dx = \frac{(x^2+1)\sqrt{(2x+3)^3}}{3}$$

$$- \frac{2x\sqrt{(2x+3)^5}}{15} + \frac{2\sqrt{(2x+3)^7}}{105} + c$$

INTEGRATION BY PARTS OF A SINGLE FUNCTION

When this method is used to integrate a single function, rather than a product of two functions, we begin by taking the function to be integrated as $\psi(x)$ in (3.2) and putting $\phi(x)$ equal to unity. Very useful applications of this method of integrating a single function are given in Example 3.3.

Example 3.3   Find

(a) $\int(\log_e x)\ dx$ (b) $\int\{\log_e(a+bx)\}\ dx$

(a) In the formula for integration by parts (3.2), put $\psi(x)=\log_e x$ and $\phi(x)=1$. Now (3.2) becomes

$$\int \log_e x\ dx = (\log_e x)\int dx - \int \left\{\frac{d(\log_e x)}{dx}\int dx\right\} dx + c$$
(3.7)

Remembering that $\int dx = x$ and $d(\log_e x)/dx = 1/x$ (3.7) becomes

$$\int (\log_e x)\ dx = x\log_e x - \int (1/x)x\ dx + c$$

$$= x\log_e x - \int dx + c$$

$$= x\log_e x - x + c$$

84

Thus

$$\int (\log_e x) \, dx = x(\log_e x - 1) + c \qquad (3.8)$$

(b)   In (3.2), put $\psi(x) = \log_e(a + bx)$ and $\phi(x) = 1$; then

$$\int \{\log_e(a + bx)\} \, dx = \log_e(a + bx) \int dx$$

$$- \int \left[\frac{d\{\log_e(a + bx)\}}{dx} \int dx \right] dx + c \qquad (3.9)$$

Now

$$\int dx = x \text{ and } \frac{d\{\log_e(a + bx)\}}{dx} = \frac{b}{(a + bx)}$$

so (3.9) becomes

$$\int \{\log_e(a + bx)\} \, dx = x \log_e(a + bx) - \int \frac{bx \, dx}{a + bx} + c \qquad (3.10)$$

The remaining integral, on the right-hand side, will have to be found by the method of substitution. Put

$$u = a + bx$$

then

$$x = \frac{(u - a)}{b}$$

and

$$dx = \frac{du}{b}$$

So

$$\int \frac{bx \, dx}{a + bx} = \int \frac{b}{u} \frac{(u - a)}{b} \frac{du}{b} = \int \frac{(u - a) \, du}{bu}$$

$$= \frac{1}{b} \int \left(1 - \frac{a}{u}\right) du = \frac{1}{b} (u - a \log_e u) + c$$

Hence, on substituting back for u,

$$\int \frac{bx \, dx}{a + bx} = \frac{1}{b} \{(a + bx) - a \log_e(a + bx)\} + c \qquad (3.11)$$

Finally, substituting (3.11) into (3.10), we have

$$\int \{\log_e(a+bx)\}\,\mathrm{d}x = x\log_e(a+bx) - \frac{1}{b}\{(a+bx)$$

$$- a\log_e(a+bx)\} + c \qquad (3.12)$$

Infinite integrals

Sometimes a definite integral may be evaluated even if one or both of the limits of integrand are infinite. The usual definition (Causton 1983, eqn 7.25) is modified slightly, and the modification differs according to which limit is infinite. Thus

$$\int_a^\infty f(x)\,\mathrm{d}x = \lim_{x\to\infty}\{F(x) - F(a)\} \qquad (3.13)$$

$$\int_{-\infty}^b f(x)\,\mathrm{d}x = \lim_{x\to-\infty}\{F(b) - F(x)\} \qquad (3.14)$$

$$\int_{-\infty}^\infty f(x)\,\mathrm{d}x = \lim_{x\to\infty}\{F(x)\} - \lim_{x\to-\infty}\{F(x)\} \qquad (3.15)$$

where

$$\frac{\mathrm{d}\{F(x)\}}{\mathrm{d}x} = f(x)$$

It should be noted that if the limit does not exist, then the integral is meaningless and cannot be evaluated.

Example 3.4   Evaluate, if possible,

(a) $\displaystyle\int_1^\infty \frac{\mathrm{d}x}{x^2}$ *(b)* $\displaystyle\int_1^\infty \frac{\mathrm{d}x}{x}$

(a)   Using (3.13), we have

$$\int_1^\infty \frac{\mathrm{d}x}{x^2} = \int_1^\infty x^{-2}\,\mathrm{d}x = \left[-x^{-1}\right]_1^\infty = \left[-\frac{1}{x}\right]_1^\infty$$

$$= \lim_{x\to\infty}\left\{\left(-\frac{1}{x}\right) - (-1)\right\} = 1$$

(b) Again using (3.13), we have

$$\int_1^\infty \frac{dx}{x} = \left[\log_e x\right]_1^\infty$$

$$= \lim_{x \to \infty} \{\log_e x - \log_e 1\}$$

Now as $x \to \infty$, $\log_e x \to \infty$. Hence the limit does not exist, and $\int_1^\infty dx/x$ is meaningless.

Application of infinite integrals: the basic mathematics of probability densities

At the beginning of Chapter 2 the equation of the probability density curve of the normal distribution was presented, and the curve itself is shown in Figure 2.1a. Although the normal distribution is the best known distribution and very often provides an adequate description of biological measurements, other measurements or phenomena are far from normally distributed and require other probability densities to describe their variability. To be eligible for a probability density, a function, $f(x)$, must have the following two properties:

$$f(x) \geqslant 0 \qquad -\infty < x < \infty \tag{3.16}$$

and

$$\int_{-\infty}^\infty f(x)\, dx = 1 \tag{3.17}$$

These two properties can be discussed qualitatively in relation to the normal distribution.

Equation (3.16) expresses the fact that, whatever the value of x, $f(x)$ is always positive; this means that the probability curve is always above the x-axis. Equation (3.17) shows that the total area under the curve is unity. Now, in a probability density, an area under the curve between two values of x is equal to the probability of finding a sample observation in that range of x. The further apart are the two values of x, the higher will be the probability of finding a sample observation between; but if the x-values are at opposite ends of the density curve, the probability becomes 1 (a certainty). This justifies the property expressed in (3.17) and also that expressed in (3.16), since the whole of this area – equal to unity – and hence the curve must be on one side (the positive side) of the x-axis.

Furthermore, the various population parameters are given by the following formulae:

Mean $\qquad\qquad \mu = \int_{-\infty}^\infty xf(x)\, dx \tag{3.18}$

Variance
$$\sigma^2 = \int_{-\infty}^{\infty} (x-\mu)^2 f(x)\, dx \tag{3.19}$$

Skewness
$$\beta_1 = \left\{ \int_{-\infty}^{\infty} (x-\mu)^3 f(x)\, dx \right\} \Big/ \sigma^3 \tag{3.20}$$

Kurtosis
$$\beta_2 = \left\{ \int_{-\infty}^{\infty} (x-\mu)^4 f(x)\, dx \right\} \Big/ \sigma^4 \tag{3.21}$$

To apply these formulae directly to the normal distribution is too complicated; so here we will illustrate the theory in relation to a mathematically much simpler case known as the **exponential distribution**. First, however, we must put the integrals in (3.19), (3.20), and (3.21) into a more convenient form.

Example 3.5 Expand the integrals in equations (a) (3.19), (b) (3.20), and (c) (3.21), assuming f(x) is a probability density.

(a)
$$\sigma^2 = \int_{-\infty}^{\infty} (x-\mu)^2 f(x)\, dx$$

$$= \int_{-\infty}^{\infty} (x^2 - 2\mu x + \mu^2) f(x)\, dx \quad \text{on expanding the bracket}$$

$$= \int_{-\infty}^{\infty} x^2 f(x)\, dx - 2\mu \int_{-\infty}^{\infty} x f(x)\, dx + \mu^2 \int_{-\infty}^{\infty} f(x)\, dx$$

$$= \int_{-\infty}^{\infty} x^2 f(x)\, dx - 2\mu^2 + \mu^2$$

The second term arises because the integral is μ (eqn(3.18)), and in the third term the integral is unity (eqn(3.17)). Hence

$$\sigma^2 = \int_{-\infty}^{\infty} x^2 f(x)\, dx - \mu^2 \tag{3.22}$$

(b) By identical steps it can be shown that the integral in (3.20) can be expanded to

$$\int_{-\infty}^{\infty} x^3 f(x)\, dx - 3\mu \int_{-\infty}^{\infty} x^2 f(x)\, dx + 2\mu^3 \tag{3.23}$$

(c) In the same way, the integral in (3.21) can be expanded to

$$\int_{-\infty}^{\infty} x^4 f(x)\, dx - 4\mu \int_{-\infty}^{\infty} x^3 f(x)\, dx + 6\mu^2 \int_{-\infty}^{\infty} x^2 f(x)\, dx - 3\mu^4 \tag{3.24}$$

88

The exponential distribution

If incidents occur at random in time or space, and if in a given interval of time or portion of space the number of incidents is independent of the number of incidents in any previous interval of time or other portion of space, then the random process involved is known as a **Poisson process**. Examples in which the above conditions may be approximately fulfilled are:

(a) counts of radioactive particles by a Geiger counter (there is a short 'dead' period of the instrument after the arrival of a particle, which affects the independence condition) over intervals of time;
(b) the occurrence of individuals of a species on a surface of a given extent, e.g. lichens along the length of a tree branch, caterpillars on the foliage surface of a plant, or plants on an area of ground (but there may be factors present to make these distributions non-random).

Let λ be the average number of incidents in a unit time interval; then the probability density of x, the duration from some arbitrary starting time to the time of the first incident (waiting time) is given by

$$f(x) = \lambda e^{-\lambda x} \qquad x, \lambda > 0 \tag{3.25}$$

The distribution has a single parameter, λ, which must be positive since it is the average number of incidents per unit of time. The time elapsing from some given time to the time of the first incident, x, must also be positive

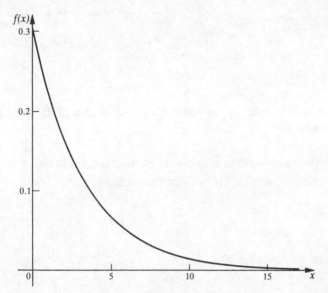

Figure 3.1 The exponential probability density $f(x) = 0.3 e^{-0.3x}\, \mathrm{d}x$.

89

since time does not move backwards. Hence the left-hand end of the curve of distribution starts at $x = 0$; but there is no limit on the upper value x may take – it is merely that large values are very unlikely (Fig. 3.1). So any integral associated with (3.25) will have limits 0 and ∞, since it is over this range that x is defined – its probability density elsewhere is zero.

The proof that (3.25) is the probability density of the stated event involves probability theory and is beyond the scope of this book. However, we may check that (3.25) is a valid probability density by seeing whether it accords with the criteria in (3.16) and (3.17). Regarding the former,

$$f(x) = \lambda e^{-\lambda x} > 0 \qquad \lambda > 0, \quad 0 \leqslant x < \infty \tag{3.26}$$

which satisfies criterion (3.16). Next we must prove that

$$\int_0^\infty \lambda e^{-\lambda x} \, dx = 1 \tag{3.27}$$

We have that

$$\int_0^\infty \lambda e^{-\lambda x} \, dx = \lambda \int_0^\infty e^{-\lambda x} \, dx$$

$$= \lambda \left[-\frac{1}{\lambda} e^{-\lambda x} \right]_0^\infty \quad \text{on integration}$$

$$= -[e^{-\lambda x}]_0^\infty \quad \text{on cancelling the } \lambda\text{s}$$

$$= -[\lim_{x \to \infty}(e^{-\lambda x}) - 1] \quad \text{on applying (3.13)}$$

$$= 1$$

since the limit is zero; so the result is proved. Hence (3.25) is a valid probability density.

POPULATION PARAMETERS OF THE EXPONENTIAL DISTRIBUTION
For the mean, from (3.18), we need to evaluate

$$\mu = \int_0^\infty x(\lambda e^{-\lambda x}) \, dx$$

which can be rewritten as

$$\mu = \lambda \int_0^\infty x e^{-\lambda x} \, dx$$

Integrating by parts (eqn(3.2)) by letting $\psi(x) = x$ and $\phi(x) = e^{-\lambda x}$, we have

$$\int xe^{-\lambda x}\,dx = x \int e^{-\lambda x}\,dx - \int \left\{ 1 \int e^{-\lambda x}\,dx \right\} dx$$

$$= -\frac{1}{\lambda} xe^{-\lambda x} - \int -\frac{1}{\lambda} e^{-\lambda x}\,dx$$

$$= -\frac{1}{\lambda} xe^{-\lambda x} + \frac{1}{\lambda} \int e^{-\lambda x}\,dx$$

$$= -\frac{1}{\lambda} xe^{-\lambda x} - \frac{1}{\lambda^2} e^{-\lambda x}$$

$$= -\frac{1}{\lambda} \left(xe^{-\lambda x} + \frac{1}{\lambda} e^{-\lambda x} \right)$$

Thus

$$\lambda \int_0^\infty xe^{-\lambda x} = -\left[xe^{-\lambda x} + \frac{1}{\lambda} e^{-\lambda x} \right]_0^\infty$$

$$= -\left[\lim_{x \to \infty}(xe^{-\lambda x}) + \lim_{x \to \infty}\left(\frac{1}{\lambda} e^{-\lambda x} \right) - 0 - \frac{1}{\lambda} \right]$$

The two limits are zero; hence

$$\mu = 1/\lambda \tag{3.28}$$

Thus the mean waiting time to the first incident is the inverse of the mean number of incidents in unit time, as one would expect.

To evaluate the variance, we employ (3.22). In our present terms, we have

$$\sigma^2 = \int_0^\infty x^2 (\lambda e^{-\lambda x})\,dx - \frac{1}{\lambda^2}$$

which can be rewritten as

$$\sigma^2 = \lambda \int_0^\infty x^2 e^{-\lambda x}\,dx - \frac{1}{\lambda^2}$$

The derivation will be given in outline only. Integrating by parts, we have

$$\int x^2 e^{-\lambda x}\,dx = x^2 \int e^{-\lambda x}\,dx - \int \left\{ 2x \int e^{-\lambda x}\,dx \right\} dx$$

$$= -\frac{1}{\lambda} x^2 e^{-\lambda x} + \frac{2}{\lambda} \int xe^{-\lambda x}\,dx$$

$$= -\frac{1}{\lambda} x^2 e^{-\lambda x} - \frac{2}{\lambda^2} \left(xe^{-\lambda x} + \frac{1}{\lambda} e^{-\lambda x} \right)$$

91

The last integral has been evaluated above when dealing with the mean. Hence

$$\lambda \int_0^\infty x^2 e^{-\lambda x} \, dx = -\left[x^2 e^{-\lambda x} + \frac{2}{\lambda} \, x e^{-\lambda x} + \frac{2}{\lambda^2} \, e^{-\lambda x} \right]_0^\infty$$

$$= 2/\lambda^2$$

and so

$$\sigma^2 = 2/\lambda^2 - 1/\lambda^2 = 1/\lambda^2 \tag{3.29}$$

The variance is entirely dependent on the mean, and is in fact the square of the mean; so the larger the mean the greater the variability of the time to first incident.

The skewness is evaluated using relationship (3.23) and then putting the result into (3.20). Of the three terms in (3.23) we only need to deal with the first; the others having already been evaluated. In the present context the integral is

$$\lambda \int_0^\infty x^3 e^{-\lambda x} \, dx$$

So, in outline, integrating by parts gives

$$\int x^3 e^{-\lambda x} \, dx = x^3 \int e^{-\lambda x} \, dx - \int \left\{ 3x^2 \int e^{-\lambda x} \, dx \right\} dx$$

$$= -\frac{1}{\lambda} \, x^3 e^{-\lambda x} + \frac{3}{\lambda} \int x^2 e^{-\lambda x} \, dx$$

$$= -\frac{1}{\lambda} \left(x^3 e^{-\lambda x} + \frac{3}{\lambda} \, x^2 e^{-\lambda x} + \frac{6}{\lambda^2} \, x e^{-\lambda x} + \frac{6}{\lambda^3} \, e^{-\lambda x} \right)$$

using the result from the paragraph evaluating the variance for the last integral above. Hence

$$\lambda \int_0^\infty x^3 e^{-\lambda x} \, dx = -\left[x^3 e^{-\lambda x} + \frac{3}{\lambda} \, x^2 e^{-\lambda x} + \frac{6}{\lambda^2} \, x e^{-\lambda x} + \frac{6}{\lambda^3} \, e^{-\lambda x} \right]_0^\infty$$

$$= 6/\lambda^3$$

This is the value of the first term in (3.23). The integral in the second term of (3.23) has already been evaluated as $2/\lambda^2$. When this is multiplied by $3\mu (= 3/\lambda)$ we obtain $6/\lambda^3$; so the sum of the first two terms of (3.23) is zero. The last term is $2/\lambda^3$, which is thus the value of the whole expression (3.23). To obtain the skewness term we need to divide by σ^3, which is $1/\lambda^3$; thus

$$\beta_1 = \frac{2}{\lambda^3} \bigg/ \frac{1}{\lambda^3} = 2 \tag{3.30}$$

92

Thus the degree of skewness is independent of the parameter of the distribution and is constant for any exponential distribution.

Finally, we have the kurtosis, and we use relationship (3.24). The only new integral to be evaluated is the first one which, in the present context, is

$$\lambda \int_0^\infty x^4 e^{-\lambda x} \, dx = -\left[x^4 e^{-\lambda x} + \frac{4}{\lambda} x^3 e^{-\lambda x} + \frac{12}{\lambda^2} x^2 e^{-\lambda x} + \frac{24}{\lambda^3} xe^{-\lambda x} + \frac{24}{\lambda^4} e^{-\lambda x} \right]_0^\infty$$

$$= 24/\lambda^4$$

The complete relationship (3.24) is

$$\frac{24}{\lambda^4} - 4\mu \frac{6}{\lambda^3} + 6\mu \frac{2}{\lambda^2} - 3\mu^4$$

Substituting for $\mu(=1/\lambda)$, the above becomes

$$\frac{24}{\lambda^4} - \frac{24}{\lambda^4} + \frac{12}{\lambda^4} - \frac{3}{\lambda^4} = \frac{9}{4}$$

Hence

$$\beta_2 = \frac{9/\lambda^4}{1/\lambda^4} = 9 \tag{3.31}$$

So, as with the skewness, the kirtosis is independent of the parameter of the distribution.

Example 3.6   An exponential distribution has parameter $\lambda = 0.3$. What is the mean, variance, skewness, and kurtosis of this distribution? Also what is the probability that $4 \leqslant x \leqslant 5$?

The density is shown in Figure 3.1. From equations (3.28) to (3.31), we have

$$\mu = 3.\dot{3}$$

$$\sigma^2 = 11.\dot{1}$$

$$\beta_1 = 2$$

$$\beta_2 = 9$$

For the second part, the required probability is given by

$$p(4 \leqslant x \leqslant 5) = \int_4^5 0.3e^{-0.3x} \, dx$$

$$= -[e^{-0.3x}]_4^5$$

$$= -[e^{-1.5} - e^{-1.2}] = -(0.2231 - 0.3012)$$

Hence

$$p(4 \leqslant x \leqslant 5) = 0.0781$$

Functions defined by integrals

The gamma function

If n is positive, the infinite integral $\int_0^\infty x^{n-1}e^{-x}\,dx$ has a finite value. It is a function of n and is called the **gamma function**; we write

$$\Gamma(n) = \int_0^\infty x^{n-1}e^{-x}\,dx \qquad (3.32)$$

Γ being the capital greek letter gamma.† When $n = 1$

$$\int_0^\infty e^{-x}\,dx = -\left[e^{-x}\right]_0^\infty$$

$$= -[0-1] = 1$$

So we have

$$\Gamma(1) = 1 \qquad (3.33)$$

If now we integrate (3.32) by parts we get

$$\int_0^\infty x^{n-1}e^{-x}\,dx = x^{n-1}\int_0^\infty e^{-x}\,dx - \int_0^\infty \left[\frac{d(x^{n-1})}{dx}\int e^{-x}\,dx\right]dx$$

$$= x^{n-1}\int_0^\infty e^{-x}\,dx - \int_0^\infty \left[(n-1)x^{n-2}\int e^{-x}\,dx\right]dx$$

$$= -\left[x^{n-1}e^{-x}\right]_0^\infty + (n-1)\int_0^\infty x^{n-2}e^{-x}\,dx$$

$$= \qquad (n-1)\Gamma(n-1)$$

(see footnote on this page) since $\left[x^{n-1}e^{-x}\right]_0^\infty = 0$. Thus

$$\Gamma(n) = (n-1)\Gamma(n-1) \qquad (3.34)$$

This is an important **recurrence relation**; by combining (3.33) and (3.34) we

---

† The term n in the (n) part of the symbol $\Gamma(n)$ is known as the **argument** of Γ: the argument is one plus the exponent of x in (3.32).

may evaluate the gamma function of any integer:

$$\Gamma(2) = 1\Gamma(1) = 1$$
$$\Gamma(3) = 2\Gamma(2) = 2$$
$$\Gamma(4) = 3\Gamma(3) = 6$$
$$\Gamma(5) = 4\Gamma(4) = 24$$
$$\Gamma(6) = 5\Gamma(5) = 120$$

We also recall that in the factorial notation

$$0! = 1$$
$$1! = 1$$
$$2! = 2$$
$$3! = 6$$
$$4! = 24$$
$$5! = 120$$

and so

$$\Gamma(n) = (n-1)! \qquad n \geqslant 1 \tag{3.35}$$

We now see that for positive integers, the factorial of the integer is equivalent to the gamma function of one plus the integer. Clearly the idea of the gamma function is the more fundamental, and we now have the explanation for the curious result of $0! = 1$.

In theory we can also define gamma functions of numbers other than integers. A particularly important one for statistical distribution theory is $\Gamma(\frac{1}{2})$, but we cannot evaluate this directly from the above theory; we require another function defined by an integral, which is related to the gamma function.

Before we leave this section, we shall consider another form of the gamma function. Put $x = u^2$; then by differentiation $dx = 2u\,du$. Substituting in (3.32) gives

$$\Gamma(n) = 2 \int_0^\infty u^{2(n-1)} e^{-u^2} u\,du$$

i.e.

$$\Gamma(n) = 2 \int_0^\infty u^{2n-1} e^{-u^2}\,du \tag{3.36}$$

This will be useful in evaluating $\Gamma(\frac{1}{2})$, which we see from (3.36) is given by

$$\Gamma(\tfrac{1}{2}) = 2 \int_0^\infty e^{-u^2}\,du \tag{3.37}$$

This integral cannot, however, be evaluated by any of the methods previously considered.

The beta function

If m and n are positive, the integral $\int_0^1 x^{m-1}(1-x)^{n-1} \, dx$ is finite and is a function of both m and n. The integral is called the **beta function** and is written

$$B(m,n) = \int_0^1 x^{m-1}(1-x)^{n-1} \, dx \tag{3.38}$$

B being the *capital* greek letter beta.

First we must show that the function is symmetrical in m and n, that is, $B(m,n) = B(n,m)$. Put $u = 1 - x$; then by differentiation $du = -dx$. So we also have $x = 1 - u$ and $dx = -du$. Substituting into (3.38) gives

$$B(m,n) = -\int_1^0 (1-u)^{m-1} u^{n-1} \, du$$

There is a negative sign outside the integral; if we remove this, we must reverse the order of the limits. We then obtain

$$B(m,n) = \int_0^1 u^{n-1}(1-u)^{m-1} \, du = B(n,m)$$

and so the symmetry of the function in m and n is proved.

As with the gamma function, so may the beta function be put in a useful alternative form. This involves the use of trigonometric functions, to be introduced in the next chapter. Put $x = \sin^2 u$; then $dx = 2 \sin u \cos u \, du$ (see Example 4.2a, page 134), and substituting into (3.38) gives

$$B(m,n) = 2 \int_0^{\pi/2} (\sin^{2(m-1)} u)(\cos^{2(n-1)} u)(\sin u \cos u \, du)$$

The middle bracket in the integral follows from the identity (4.3); and the limits follow because when $x = 1$, $\sin^2 u = \sin u = 1$, which means $u = 90°$ or $\pi/2$ radians, and when $x = 0$, $\sin^2 u = \sin u = 0$, which implies $u = 0°$ or 0 radians. Hence

$$B(m,n) = 2 \int_0^{\pi/2} (\sin^{2m-1} u)(\cos^{2n-1} u) \, du \tag{3.39}$$

We see that

$$B(\tfrac{1}{2}, \tfrac{1}{2}) = 2 \int_0^{\pi/2} du = 2[u]_0^{\pi/2} = \pi \tag{3.40}$$

The relationship between gamma and beta functions

It can be shown that

$$B(m, n) = \frac{\Gamma(m)\,\Gamma(n)}{\Gamma(m+n)} \qquad (3.41)$$

Besides being useful in its own right, (3.41) also provides two useful recurrence relationships for calculating values of the beta function. For

$$B(m+1, n) = \frac{\Gamma(m+1)\,\Gamma(n)}{\Gamma(m+n+1)}$$

$$= \frac{m\Gamma(m)\,\Gamma(n)}{(m+n)\,\Gamma(m+n)}$$

$$= \frac{m}{m+n}\,\frac{\Gamma(m)\,\Gamma(n)}{\Gamma(m+n)}$$

i.e.

$$B(m+1, n) = \frac{m}{m+n}\,B(m, n) \qquad (3.42)$$

Because of the symmetry, a similar expression obtains for $B(m, n+1)$. There is a similar rule for the double increment:

$$B(m+1, n+1) = \frac{\Gamma(m+1)\,\Gamma(n+1)}{\Gamma(m+n+2)}$$

$$= \frac{m\Gamma(m)\,n\Gamma(n)}{(m+n+1)\,\Gamma(m+n+1)} \qquad \text{by 3.34}$$

$$= \frac{mn\Gamma(m)\,\Gamma(n)}{(m+n+1)(m+n)\,\Gamma(m+n)}$$

and so $B(m+1, n+1) = \dfrac{mn}{(m+n+1)(m+n)}\,B(m, n) \qquad (3.43)$

Applying (3.42) and (3.43), we see from (3.38) that

$$B(1, 1) = \int_0^1 dx = [x]_0^1 = 1$$

so

$$B(2, 1) = \tfrac{1}{2} \times 1 = \tfrac{1}{2}$$

and

$$B(2, 2) = \frac{1}{(3)(2)}\,B(1, 1) = \tfrac{1}{6}$$

This is *much* easier than evaluating (3.38) directly.

The evaluation of $\displaystyle\int_0^\infty e^{-x^2}\,dx$

From (3.41) we have that

$$B(\tfrac{1}{2},\tfrac{1}{2}) = \frac{\Gamma(\tfrac{1}{2})\,\Gamma(\tfrac{1}{2})}{\Gamma(\tfrac{1}{2}+\tfrac{1}{2})} = \frac{\Gamma^2(\tfrac{1}{2})}{\Gamma(1)} = \Gamma^2(\tfrac{1}{2})$$

where $\Gamma^2(\tfrac{1}{2}) = \{\Gamma(\tfrac{1}{2})\}^2$. But from (3.40), $B(\tfrac{1}{2},\tfrac{1}{2}) = \pi$. Hence

$$\Gamma(\tfrac{1}{2}) = \sqrt{\pi} \tag{3.44}$$

Substituting for $\Gamma(\tfrac{1}{2})$ in (3.37) therefore yields

$$\int_0^\infty e^{-x^2}\,dx = \tfrac{1}{2}\sqrt{\pi} \tag{3.45}$$

Applications of gamma functions

The normal distribution

The probability density of the normal distribution was given in (2.1). If we make the substitution $z = (x - \mu)/\sigma$, then z may be regarded as a normally distributed variate with a mean of zero and standard deviation of one ($\mu = 0$, $\sigma = 1$). In considering the probability density function of z, we substitute z for x in (2.1) and also bear in mind that $\mu = 0$ and $\sigma = 1$:

$$f(z) = \frac{1}{\sqrt{(2\pi)}}\, e^{-\frac{1}{2}z^2} \tag{3.46}$$

Equation (3.46) is the probability density of what is known as the **standard normal distribution,** and z is called the **standard normal deviate.** Evidently $f(z)$ must be positive for any value of $z(-\infty < z < \infty)$, so property (3.16) is satisfied for (3.46) to be a probability density. It now remains to deal with criterion (3.17), namely, to show that

$$\frac{1}{\sqrt{(2\pi)}} \int_{-\infty}^{\infty} e^{-\frac{1}{2}z^2}\,dz = 1 \tag{3.47}$$

To prove (3.47) we need to show that

$$\int_{-\infty}^{\infty} e^{-\frac{1}{2}z^2}\,dz = \sqrt{(2\pi)} \tag{3.48}$$

Now we already have the result that $\Gamma(\tfrac{1}{2}) = \sqrt{\pi}$ (eqn(3.44)), i.e. that

$$\int_0^\infty x^{-1/2}e^{-x}\,dx = \sqrt{\pi} \tag{3.49}$$

Now make the substitution $x = \frac{1}{2}z^2$; then $\mathrm{d}x = z\,\mathrm{d}z$; so

$$\int_0^\infty x^{-\frac{1}{2}}e^{-x}\,\mathrm{d}x = \int_0^\infty (\tfrac{1}{2}z^2)^{-1/2}(e^{-\frac{1}{2}z^2})z\,\mathrm{d}z$$

$$= \frac{1}{2^{-1/2}}\int_0^\infty z^{-1}e^{-\frac{1}{2}z^2}z\,\mathrm{d}z$$

i.e.

$$\int_0^\infty x^{-1/2}e^{-x}\,\mathrm{d}x = \sqrt{2}\int_0^\infty e^{-\frac{1}{2}z^2}\,\mathrm{d}z$$

Substituting for the integral on the left-hand side, using (3.49), gives, after rearrangement,

$$\int_0^\infty e^{-\frac{1}{2}z^2}\,\mathrm{d}z = \frac{\sqrt{\pi}}{\sqrt{2}} \tag{3.50}$$

But since the density of z is symmetrical about $z = 0$, the area on the left-hand side ($z < 0$) is the same as that on the right ($z > 0$), which is given by (3.50). Therefore

$$\int_{-\infty}^\infty e^{-\frac{1}{2}z^2}\,\mathrm{d}z = 2\int_0^\infty e^{-\frac{1}{2}z^2}\,\mathrm{d}z = \frac{2\sqrt{\pi}}{\sqrt{2}} = \sqrt{(2\pi)}$$

and the required result, (3.48), is proved.

A model for calculating tree stem volume

In estimating the amount of marketable timber contained in a single tree of a stand, the forester requires not only an estimate of the total volume of the tree trunk from linear measurements of height and basal diameter (actually *diameter at breast height*, 1.5 m above ground) but also some information on the shape of the stem in order to determine the length of a plank of given width that is obtainable. Total timber volume estimates may be obtained from height and diameter measurements by an allometric equation (see Causton 1983, p. 51), but estimating the amount of *merchantable* volume requires a geometric model of the tree trunk. Very many such models have been proposed, most of them based on solids of revolution (page 112) of low degree polynomials, or on simple geometric solids such as the cone, paraboloid, neiloid,† or cylinder.

As might be expected, none of these wholly empirical models provides a really satisfactory description of the geometry of a tree trunk; a mechanistic input is required for greater realism (see page 246 for a discussion on the nature of empirical and mechanistic models). Recently, Forslund (1982)

---

†A neiloid is the solid of revolution of Neil's parabola: $y = ax^{3/2}$.

proposed a geometrical tree model volume starting from a general, but empirical, stem profile equation and employing the position of the centre of gravity of the trunk to enable the shape to be defined for particular trees.

DERIVATION OF THE STEM VOLUME EQUATION

The general profile equation relating diameter and height is

$$d = D\{1 - (h/H)^b\}^{1/a} \tag{3.51}$$

where d is stem diameter at height h, D is basal diameter, and H is total stem height. It is mathematically more convenient to work with the radius, $r(=d/2)$, and so (3.51) becomes

$$r = D\{1 - (h/H)^b\}^{1/a}/2 \tag{3.52}$$

and the profile curve of this equation is shown in Figure 3.2. The volume of the solid of revolution of this relationship is given by

$$V = \pi \int_0^H r^2 \, dh$$

and so, substituting equation (3.52) for r, we have

$$V = \frac{\pi D^2}{4} \int_0^H \{1 - (h/H)^b\}^{2/a} \, dh \tag{3.53}$$

Now let $K = h/H$, that is K is the relative height at h and lies in the range 0 (at the base of the tree) and 1 (at the top); therefore $dK/dh = 1/H$, i.e. $dh = H \, dK$. Changing the variable in (3.53) yields

$$V = \frac{\pi D^2 H}{4} \int_0^1 (1 - K^b)^{2/a} \, dK \tag{3.54}$$

Note the corresponding change in the limits of integration. If we now let

$$y = K^b \tag{3.55}$$

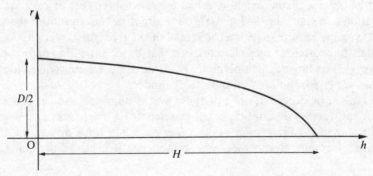

Figure 3.2 The curve of the equation $r = D[1 - (h/H)^b]^{1/a}/2$.

then $dy/dK = bK^{b-1}$, i.e. $dK = dy/(bK^{b-1}) = K^{1-b} \, dy/b$. From (3.55) $K = y^{1/b}$ and $K^{-b} = y^{-1}$; hence

$$dK = \frac{y^{1/b-1} \, dy}{b} \qquad (3.56)$$

Substituting (3.55) and (3.56) into (3.54) gives

$$V = \frac{\pi D^2 H}{4} \int_0^1 (1-y)^{2/a} \frac{y^{1/b-1} \, dy}{b}$$

i.e.

$$V = \frac{\pi D^2 H}{4b} \int_0^1 y^{1/b-1}(1-y)^{2/a} \, dy \qquad (3.57)$$

Equation (3.57) contains a beta function. On comparing (3.57) with (3.38), we can write the former as

$$V = \frac{\pi D^2 H}{4b} B(1/b, 2/a + 1) \qquad (3.58)$$

Further, using (3.41), we find that

$$V = \frac{\pi D^2 H}{4b} \cdot \frac{\Gamma(2/a+1) \, \Gamma(1/b)}{(2/a + 1/b + 1)} \qquad (3.59)$$

Now from equation (3.51) we have that

$$D^2 = d^2 \left(\frac{1}{1-K^b}\right)^{2/a}$$

and substituting in (3.59) gives

$$V = \frac{\pi d^2 H}{4} \left(\frac{1}{1-K^b}\right)^{2/a} \cdot \frac{1}{b} \cdot \frac{\Gamma(2/a+1) \, \Gamma(1/b)}{\Gamma(2/a + 1/b + 1)}$$

Finally, since $(1/b) \, \Gamma(1/b) = \Gamma(1/b + 1)$ (cf.(3.34)), we have

$$V = \frac{\Gamma(2/a + 1) \, \Gamma(1/b + 1)}{\Gamma(2/a + 1/b + 1)} \left(\frac{1}{1-K^b}\right)^{2/a} \frac{\pi d^2 H}{4} \qquad (3.60)$$

Equation (3.60) gives the volume of tree trunk above any height where the diameter is d.

CENTRE OF GRAVITY

To define the position (height, $\bar{h}$) of the centre of gravity, we first note that for the volume of the solid of revolution defined by equation (3.51), we have

$$\bar{h} = M/V \qquad (3.61)$$

101

where V is the volume and M is the first moment about the origin given by

$$M = \frac{\pi}{4} \int_0^H hd^2 \, dh \tag{3.62}$$

The derivation now proceeds in a similar way to the volume equation above, and is left as an exercise for you (see Exercise 3.7 at the end of this chapter). The result is

$$\bar{K} = \frac{\Gamma(2/b) \cdot \Gamma(2/a + 1/b + 1)}{\Gamma(1/b) \cdot \Gamma(2/a + 2/b + 1)} \tag{3.63}$$

USE OF THE MODEL

Since the centre of gravity was the only physical property of the tree stem assessed, the value of one of the two parameters needs to be fixed. By putting $b = 1$, (3.51) becomes

$$d = D(1 - h/H)^{1/a} \tag{3.64}$$

which, depending on the value of a, includes the conic sections (page 197) commonly used to describe tree form. Putting $b = 1$ in (3.60) gives

$$V = \frac{\Gamma(2/a + 1) \, \Gamma(2)}{\Gamma(2/a + 2)} \left(\frac{1}{1 - K}\right)^{2/a} \frac{\pi d^2 H}{4} \tag{3.65}$$

Now $\Gamma(2) = 1$ (page 95), and from (3.34) we have

$$\frac{\Gamma(n - 1)}{\Gamma(n)} = \frac{1}{n - 1}$$

so putting $n = 2/a + 2$, (3.65) becomes

$$V = \left(\frac{a}{a + 2}\right) \left(\frac{1}{1 - K}\right)^{2/a} \frac{\pi d^2 H}{4} \tag{3.66}$$

Substituting $b = 1$ into (3.63) gives

$$\bar{K} = \frac{\Gamma(2) \, \Gamma(2/a + 2)}{\Gamma(1) \, \Gamma(2/a + 3)}$$

and by an analogous argument to the above, we get

$$\bar{K} = \frac{a}{2(a + 1)} \tag{3.67}$$

By changing the value of a a continuum of forms results, including the following familiar solids: neiloid ($a = 0.6$, $\bar{K} = 0.2$); cone ($a = 1$, $\bar{K} = 0.25$); paraboloid ($a = 2$, $\bar{K} = 0.3$); cylinder (a infinitely large, $\bar{K} = 0.5$). The $\bar{K}$-values are obtained from (3.67); and the last value, for the cylinder, is obtained by the limit

$$\bar{K} = \lim_{a \to \infty} \left\{ \frac{a}{2(a + 1)} \right\}$$

A numerical application of the model, using Forslund's data, is given as an exercise at the end of the chapter (Exercise 3.8).

Multiple integrals

Hitherto we have been concerned with integrals of functions of one variable, the geometrical interpretation of such functions being curves in two-dimensional space. In particular, we have found that the area under the curve of a function, i.e. the area bounded by the curve, the horizontal or x-axis, and two vertical lines at $x = a$ and $x = b$, is given by the definite integral of the function between the limits of a and b. We shall now consider, in a very preliminary way only since the subject is difficult, the situation involving a function of two variables; the corresponding problem is that of finding the volume under a surface. Before embarking on this section, you should first read Chapter 6, which deals with the background mathematics and geometrical considerations involved in functions of more than one variable. The material below is based on the concepts and notation of Chapter 6.

Constant limits in double integrals

A function of the form $z = f(x, y)$ is geometrically represented as a surface in three-dimensional space in which the x–y axial plane is horizontal and the z-axis is vertical. The volume under the surface of the function is the volume between the surface, the horizontal or x–y axial plane, and some kind of vertical circumscribing surface. In the simplest case, the vertical circumscribing surface consists of four vertical planes, two of which are parallel to the x–z axial plane and two parallel to the y–z axial plane, and so these planes intersect vertically at right angles as in a box. The surface representing $f(x, y)$ thus forms the 'lid' of the box; but the tops of the sides may be very irregular, depending on the nature of the surface $f(x, y)$. The box intersects the x–y horizontal surface as a rectangle (EFGH, Fig. 3.3). Let the lengths of the sides of the rectangle on the x–y plane be $b - a$ in the x-direction and $d - c$ in the y-direction; and let the rectangle be dissected into a series of very thin strips, of width δx, by lines parallel to the y-axis. A typical strip is shown in Figure 3.3 whose left-hand edge, IJ, is distant x from the origin.

Now above IJ can be imagined a vertical plane (Fig. 3.4) whose base is

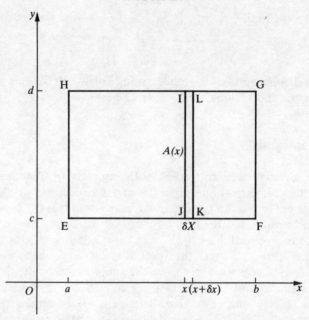

Figure 3.3 The rectangular area of the limits of integration, on the x–y plane, for the double integral $\int_a^b\int_c^d f(x,y)\,dy\,dx$.

IJ, whose sides, JM and IN, lie on the vertical circumscribing surfaces based on EF and GH, and whose top, MN, is the curve of the surface $f(x, y)$ in the y-direction at the fixed value of x. Clearly the area of this vertical surface will depend on x because if a different x-value (different strip) is chosen, although the length of the base will remain the same $(d - c)$ the lengths of the vertical sides will change with the conformation of the surface $f(x, y)$ and so therefore will the area of the vertical surface, denoted a $A(x)$. The volume of the element above IJKL is then approximately $A(x)\,\delta x$. As $\delta x \to 0$ (the strips become more numerous) the volume of each element approaches $A(x)\,\delta x$ more exactly; and when δx becomes vanishingly small, the total volume under the surface between the limits of $x = a$ and $x = b$ (and, of course, between $y = c$ and $y = d$, already implied is given by

$$V = \int_a^b A(x)\,dx \qquad (3.68)$$

But for any x-value, $A(x)$ is given by

$$A(x) = \int_c^d f(x, y)\,dy \qquad (3.69)$$

104

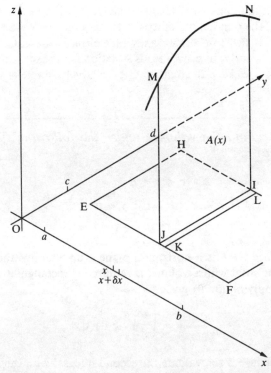

Figure 3.4 The $A(x) = \int_c^d f(x, y)\mathrm{d}y$ component of the double integral $\int_a^b\int_c^d f(x, y)$ $\mathrm{d}y\,\mathrm{d}x$.

where x is constant. Putting (3.68) and (3.69) together, we have

$$V = \int_a^b \left\{ \int_c^d f(x, y)\;\mathrm{d}y \right\}\;\mathrm{d}x$$

Notice that the latter variable of integration, $\mathrm{d}x$, is associated with the first integral sign with its limits; using this convention we may write

$$V = \int_a^b \int_c^d f(x, y)\;\mathrm{d}y\;\mathrm{d}x \qquad (3.70)$$

If strips parallel to the x-axis had been taken and the above derivation carried through, we would obtain

$$V = \int_c^d \int_a^b f(x, y)\;\mathrm{d}x\;\mathrm{d}y \qquad (3.71)$$

Equations (3.70) and (3.71) must yield the same numerical result, but the interchange of the order of integration is only as simple as this when all four

105

limits involved are constants. Geometrically, it is because the intersection of the edges of the required volume with the x-y plane gives a rectangle with sides parallel to the co-ordinate axes that ensures the constant limits of integration. Evidently, in practice this situation is the exception rather than the rule, but it does keep the mathematics relatively simple as the following example shows.

Example 3.7   Find the volumes under the following surfaces with the limits shown:

(a) $z = k;$ $a < x < b,$ $c < y < d$

(b) $z = gx + hy;$ $a < x < b,$ $c < y < d$

(c) $z = 1 - x^2 - y^2;$ $-0.7 < x, y < 0.7$

(a) The surface $z = k$ is a horizontal plane k units above the x–y plane. Hence the solid whose volume is required is rectangular in all dimensions. Applying (3.70) gives

$$V = \int_a^b \int_e^d k \; dy \; dx$$

The procedure is to evaluate the second integral with the first variable of integration and then the first integral with the second variable of integration; hence, integrating with respect to y (keeping x constant) gives

$$V = \int_a^b [ky]_c^d \; dx = k \int_a^b (d - c) \; dx$$

Now integrate with respect to x:

$$V = k[(d - c)x]_a^b = k(d - c)(b - a) \quad \text{cubic units}$$

So, as expected, the volume is the product of height, breadth, and depth.

(b) This surface slopes in the x-direction with gradient g and in the y-direction with gradient h. Hence the solid whose volume is required is similar to that in (a) but with a sloping top. The double integral giving the required volume is

$$V = \int_a^b \int_c^d (gx + hy) \; dy \; dx$$

Integrating first with respect to y, keeping x constant:

$$V = \int_a^b [\,gxy + \tfrac{1}{2}hy^2\,]_c^d \, dx = \int_a^b (gdx + \tfrac{1}{2}hd^2 - gcx - \tfrac{1}{2}hc^2) \, dx$$

$$= \int_a^b \{g(d-c)x + \tfrac{1}{2}h(d^2 - c^2)\} \, dx$$

Now integrate with respect to x:

$$V = [\tfrac{1}{2}g(d-c)x^2 + \tfrac{1}{2}h(d^2 - c^2)x\,]_a^b$$

$$= \tfrac{1}{2}\{g(d-c)b^2 - g(d-c)a^2 + h(d^2 - c^2)b - h(d^2 - c^2)a\}$$

$$= \tfrac{1}{2}\{g(d-c)(b^2 - a^2) + h(d^2 - c^2)(b - a)\} \quad \text{cubic units}$$

(c) The equation $z = 1 - x^2 - y^2$ is that of a paraboloid of the kind shown in Figure 6.6a. The axis of this paraboloid is the z-axis, and the paraboloid intersects the horizontal x–y axial plane ($z = 0$) in a circle whose equation is $x^2 + y^2 = 1$, whose centre is the origin, and whose radius is 1 unit. The apex of the paraboloid intercepts the z-axis at $z = 1$. However, we do not find the whole volume beneath the top of the paraboloid down to the horizontal plane, but only that volume beneath the paraboloid bounded by the square ABCD on the x–y plane. The corners of this square are just inside the circle of intersection by the paraboloid on the x–y plane (Fig. 3.5). The required volume is given by the double integral

$$V = \int_{-0.7}^{0.7} \int_{-0.7}^{0.7} (1 - x^2 - y^2) \, dy \, dx$$

Integrating with respect to y gives

$$V = \int_{-0.7}^{0.7} [\,(1 - x^2)y - \tfrac{1}{3}y^3\,]_{-0.7}^{0.7} \, dx$$

which results, after a number of straightforward algebraic steps, in

$$V = 1.4 \int_{-0.7}^{0.7} (0.83\dot{6} - x^2) \, dx$$

$$= 1.4\,[\,0.83\dot{6}x - \tfrac{1}{3}x^3\,]_{-0.7}^{0.7}$$

$$= 1.319\,7\dot{3} \quad \text{cubic units}$$

In the above three examples, the equations of the surfaces have been kept mathematically simple in order to demonstrate principles in as straightforward a manner as possible. Even replacing the paraboloid by a hemisphere in Example 3.7c would have involved much more difficult integrations.

107

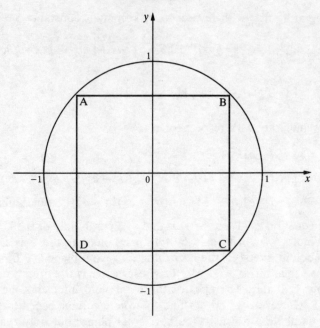

Figure 3.5 The circle $x^2 + y^2 = 1$, being the intersection of the paraboloid $z = 1 - x^2 - y^2$ with the x–y plane; and the square area of the limits of integration on the x–y plane for the double integral $\int_{-0.7}^{0.7}\int_{-0.7}^{0.7}(1 - x^2 - y^2)\,\mathrm{d}y\,\mathrm{d}x$.

Finally, as with a single definite integral, a double integral does not have to be geometrically identified with a volume. One could write quite simply that

$$\int_{-0.7}^{0.7}\int_{-0.7}^{0.7}(1 - x^2 - y^2)\,\mathrm{d}y\,\mathrm{d}x = 1.319\,7\dot{3}$$

Variable limits in double integrals

Constant limits are obviously rare in practice, and we must now consider the somewhat more difficult case where the region on the x-y plane above which the volume under the surface is required can be any shape. So long as the region is enclosed by a continuous curve which can be referred to a known mathematical function, $y = y(x)$, then the expression for the required volume can be written down and evaluated; if the resulting integrals are intractable, they may be evaluated numerically.

The situation is shown in Figure 3.6, which is analogous to that depicted in Figure 3.3 for constant limits. Again, EFGH is the region over which

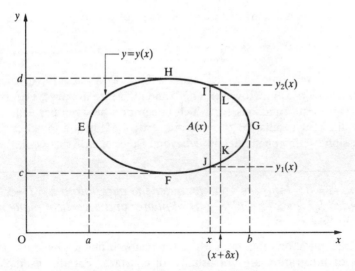

Figure 3.6 A non-rectangular area of the limits of integration, on the $x-y$ plane, for the double integral $\int_a^b \int_{y_1(x)}^{y_2(x)} f(x, y) \, dy \, dx$.

integration is required, with extremities in the x-direction of a and b and in the y-direction of c and d. Dissect the area into a large number of very thin strips of width δx parallel to the y-axis, and a typical strip IJKL whose left-hand edge is x units along the x-axis is shown in Figure 3.6. Above IJ, and bounded by vertical lines from I and J and the curve of the surface itself at that value of x, is a vertical plane of area $A(x)$. As $\delta x \to 0$, the total volume under the surface and above the region in question is given by

$$V = \int_a^b A(x) \, dx$$

So far this is all as before. The new feature is that as x changes so does the length of the base line of $A(x)$, IJ; and this length alters in accordance with the function $y = y(x)$. So the limits are obviously a function of x, and $A(x)$ is given by

$$A(x) = \int_{y_1(x)}^{y_2(x)} f(x, y) \, dy$$

where $y_1(x)$ is the value of $y(x)$ at point J, and $y_2(x)$ is the value of $y(x)$ at point I. Putting these two relationships together, we get

$$V = \int_a^b \int_{y_1(x)}^{y_2(x)} f(x, y) \, dy \, dx \qquad (3.72)$$

109

If it is more convenient to reverse the roles of x and y, we have

$$V = \int_c^d \int_{x_1(y)}^{x_2(y)} f(x, y) \, \mathrm{d}x \, \mathrm{d}y \qquad (3.73)$$

where now we assume that the region is bounded by a curve of the functional form $x = x(y)$. However, (3.72) and (3.73) are not always equivalent. There are some functions for which integration between particular limits gives different results for the different orders of integration, but further discussion of this phenomenon is beyond the scope of this book.

Example 3.8   *Find the volume under the paraboloid $z = 1 - x^2 - y^2$ bounded by the x–z and y–z axial planes and a vertical plane above the line $y = 1 - x$.*

This example shows that even where the region is defined by straight lines, limits of integration are not usually all constant. For this example, the volume is given by

$$V = \int_0^1 \int_0^{1-x} (1 - x^2 - y^2) \, \mathrm{d}y \, \mathrm{d}x$$

Integrating with respect to y gives

$$V = \int_0^1 [(1 - x^2)y - \tfrac{1}{3}y^3]_0^{1-x} \, \mathrm{d}x$$

$$= \int_0^1 \{(1 - x^2)(1 - x) - \tfrac{1}{3}(1 - x)^3\} \mathrm{d}x$$

which simplifies to

$$V = 2 \int_0^1 (\tfrac{1}{3} - x^2 + \tfrac{2}{3}x^3) \, \mathrm{d}x$$

$$= 2[\tfrac{1}{3}x - \tfrac{1}{3}x^3 + \tfrac{1}{6}x^4]_0^1$$

$$= 2(\tfrac{1}{3} - \tfrac{1}{3} + \tfrac{1}{6}) = \tfrac{1}{3} \quad \text{cubic units}$$

Example 3.9   *Find   the   volume   beneath   the   paraboloid $z = 1 - x^2 - y^2$.*

The form of the question implies that we require the whole volume beneath the paraboloid down to where it intersects the x–y plane. The curve of intersection is the circle $x^2 + y^2 = 1$ (see Example 3.7(c)). Rearranging the equation of the circle to obtain y in terms of x gives

$$y = \pm \sqrt{(1 - x^2)}$$

110

This circle is shown in Figure 3.5; x-values range between -1 and 1, and positive values of y are given by the positive square root and vice versa. Application of equation (3.72) gives in this instance

$$V = \int_{-1}^{1} \int_{-\sqrt{(1-x^2)}}^{\sqrt{(1-x^2)}} (1 - x^2 - y^2) \, dy \, dx$$

Integration with respect to y gives

$$V = \int_{-1}^{1} [(1 - x^2)y - \tfrac{1}{3}y^3]_{-\sqrt{(1-x^2)}}^{\sqrt{(1-x^2)}} \, dx$$

$$= \int_{-1}^{1} [\{(1 - x^2)^{3/2} - \tfrac{1}{3}(1 - x^2)^{3/2}\}$$

$$- \{-(1 - x^2)^{3/2} + \tfrac{1}{3}(1 - x^2)^{3/2}\}] \, dx$$

which simplifies to

$$V = 1\tfrac{1}{3} \int_{-1}^{1} (1 - x^2)^{3/2} \, dx \qquad (3.74)$$

Although this integral looks simple, it is not at all easy to integrate. Rather than presenting the method, this seems a good place to introduce published tables of integrals; several such tables are available, and a selection are detailed at the end of this chapter. For the present problem, we find that entry 206 in Weast and Astle (1979) reads

$$\int \sqrt{(a^2 - x^2)^3} \, dx = \frac{1}{4}\left\{x\sqrt{(a^2 - x^2)^3} + \frac{3a^2 x}{2} \sqrt{(a^2 - x^2)} + \frac{3a^4}{2} \sin^{-1}\left(\frac{x}{|a|}\right)\right\}$$

Substituting $a = 1$, (3.74) becomes

$$V = (\tfrac{4}{3})(\tfrac{1}{4})[x(1 - x^2)^{3/2} + \tfrac{3}{2}x(1 - x^2)^{1/2} + \tfrac{3}{2}\sin^{-1} x]_{-1}^{1}$$

On substituting the limits, we find that the first two terms within the square brackets vanish. Regarding the remaining term, the angle whose sine is 1 is $\pi/2$ radians and $\sin^{-1}(-1)$ is $-\pi/2$ radians, thus

$$V = \frac{1}{3}\left\{\frac{3}{2}\frac{\pi}{2} - \frac{3}{2}\left(-\frac{\pi}{2}\right)\right\} = \tfrac{1}{2}\pi \quad \text{cubic units}$$

Multiple integrals

The theory considered in this part of the chapter extends, without any new principles, to functions of more than two variables. Multiple integrals are important in the mathematics of electricity and magnetism and, of more immediate concern to the biologist, diffusion and heat conduction. A detailed introduction to these topics can be found in Ledermann (1966).

Volume of a solid of revolution

The method to be described in this section has nothing to do with multiple integrals, but since the preceding part of the chapter has been much concerned with volumes this alternative approach is conveniently placed here.

Consider the curve of the function $y = f(x)$ in Figure 3.7a. Imagine now that this curve is rotated rigidly about the x-axis; this will generate a solid whose upper and lower bounds in the y-direction will appear as the curve itself and its mirror image depicted as the hatched curve in Figure 3.7a. Any cross-section of the solid in the y-direction is circular. The solid generated by this method is called a **solid of revolution,** and the problem addressed here is to find its volume between the limits of $x = a$ and $x = b$.

Dissect the solid into a large number of thin discs by planes perpendicular to the x-axis. A typical disc is shown distant x along the x-axis in Figure 3.7a. The volume of this disc is approximately $A\,\delta x$, where A is the area of the plane at x. As $\delta x \to 0$, the volume of the disc approaches $A\,\delta x$ more exactly, and when δx becomes vanishingly small the total volume of the solid between $x = a$ and $x = b$ is given by

$$V = \int_a^b A \; \mathrm{d}x$$

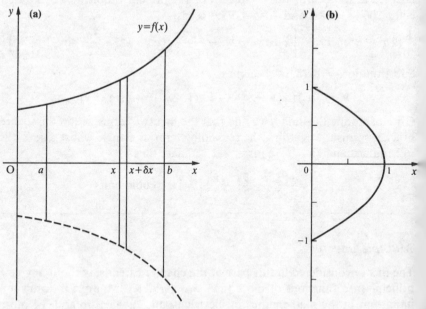

Figure 3.7 (a) The principle of the volume of a solid of revolution. (b) Curve of the function $y^2 = 1 - x$.

But $A = \pi y^2$, where y is the radius of the disc at x; hence

$$V = \pi \int_a^b y^2 \, dx \qquad (3.75)$$

or

$$V = \pi \int_a^b \{f(x)\}^2 \, dx \qquad (3.76)$$

According to the problem in hand, it may be preferable to rotate the curve around the y-axis; then

$$V = \pi \int_c^d x^2 \, dy \qquad (3.77)$$

where $y = c$ and $y = d$ are the limits of integration.

Example 3.10 Find the volume of the solid of revolution of the parabola $x = 1 - y^2$ about the x-axis between $x = 0$ and its apex.

By substitution it can be found that when $x = 0$, $y = \pm 1$, and when $y = 0$, $x = 1$. This means that the parabola intersects the y-axis at ± 1 and the x-axis at 1. Further, $dx/dy = -2y$, and so when $dx/dy = 0$, $y = 0$, which shows that the apex of the parabola is at the point $(1, 0)$. The curve is shown in Figure 3.7b. Rotation of this curve around the x-axis gives a paraboloid whose base is on the y-axis between the values ± 1 and whose apex is on the x-axis at $x = 1$; it is the identical paraboloid whose volume was calculated in Example 3.9. For the present method, rearrangement of the equation of the parabola gives $y^2 = 1 - x$; hence $\{f(x)\}^2 = 1 - x$, and so applying (3.76) yields

$$V = \pi \int_0^1 (1 - x) \, dx$$
$$= \pi [x - \tfrac{1}{2}x^2]_0^1$$
$$= \pi(1 - \tfrac{1}{2}) = \tfrac{1}{2}\pi \quad \text{cubic units}$$

as before.

Suggestions for further reading, and other references

Much of the material of this chapter is standard calculus for which many textbooks exist. For further details or more advanced topics, you are recommended to examine a number of books and use the one which appears to suit you best.

A book that I find useful, which goes a little further with many basic topics, is **Bowman, F.** (1931). *Elementary calculus*. London: Longman. (Numerous reprints.)

More modern books are:

Abbott, P. (1972). *Teach yourself calculus*. Sevenoaks: English Universities Press (now Hodder & Stoughton Educational).

Heading, J. (1970). *Mathematical methods in science and engineering*, 2nd edn. London: Edward Arnold.

Pedoe, J. (1980). *Higher mathematics*. London: Allen & Unwin.

Statistical probability densities are derived in most books on mathematical statistics. A more descriptive account of the elements of mathematical statistics, which could serve as a useful 'lead in' to the usual textbooks on this subject, has been written by **Bulmer, M. G.** (1965). *Principles of statistics*. Edinburgh: Oliver & Boyd.

Multiple integrals are dealt with clearly by **Ledermann, W.** (1966). *Multiple integrals* (Library of Mathematics series). London: Routledge & Kegan Paul.

For tables of integrals, there are the following. The first is the most accessible and contains several hundred entries; the remainder are of increasing order of size and complexity.

Weast, R. C. and M. J. Astle, (eds) (1979). *CRC handbook of chemistry and physics,* 59th edn. Boca Raton, Florida: CRC Press.

Dwight, H. B. (1934). *Tables of integrals and other mathematical data*. London: Macmillan.

Bois, G. P. (1961). *Tables of indefinite integrals*. New York: Dover Publications.

Gradhteyn, I. S. and **I. M. Ryzhik,** (1965). *Tables of integrals, series, and products,* 4th edn (translated by A. Jeffrey). New York: Academic Press.

The remaining reference in this chapter is **Forslund, R. R.** (1982). A geometrical tree volume model based on the location of the centre of gravity of the bole. *Canadian Journal of Forest Research,* **12**, 215–21.

Exercises

3.1 Using the substitutions indicated, find

(a) $\int \dfrac{dx}{x\sqrt{(1+x^2)}}$ putting $z = \sqrt{(1+x^2)}$

(b) $\int \dfrac{x\,dx}{(2x+1)^3}$ putting $z = 2x+1$

(c) $\int \dfrac{x\,dx}{1+x^2}$ putting $z = 1+x^2$

3.2 Integrate by parts:

(a) $\int xe^x \, dx$ (b) $\int x^2 e^x \, dx$ (c) $\int x(\log_e x) \, dx$

(d) $\int \{ (\log_e x)/x^2 \} \, dx$

3.3 Evaluate

(a) $\int_0^1 xe^x \, dx$ (b) $\int_{-\infty}^{\infty} \dfrac{dx}{(x^2 + 1)}$

(c) $\int_1^{\infty} \dfrac{dx}{\sqrt{x}}$ (d) $\int_0^{\infty} \dfrac{dx}{(x+2)^2}$

3.4 For the exponential distribution $f(x) = 0.3e^{-0.3x}$, find the value of $x(=l)$ such that there is a probability of 0.025 of obtaining an observation in the range $0 \leqslant x \leqslant l$ and another value of $x(=u)$ such that there is a probability of 0.025 of obtaining an observation in the range $u \leqslant x < \infty$. What is the probability of finding an observation in the range $l \leqslant x \leqslant u$?

3.5 Evaluate

(a) $\Gamma(1\tfrac{1}{2})$ (b) $\Gamma(2\tfrac{1}{2})$ (c) $\Gamma(3\tfrac{1}{2})$ (d) $\Gamma(4\tfrac{1}{2})$

3.6 Evaluate

(a) $B(1\tfrac{1}{2}, \tfrac{1}{2})$ (b) $B(1\tfrac{1}{2}, 1\tfrac{1}{2})$ (c) $B(2\tfrac{1}{2}, 1\tfrac{1}{2})$ (d) $B(2\tfrac{1}{2}, 2\tfrac{1}{2})$

3.7 Starting from equations (3.61), (3.62) and (3.58) of Forslund's geometric tree volume model, show that

$$\bar{K} = \frac{\Gamma(2/b)\,\Gamma(2/a + 1/b + 1)}{\Gamma(1/b)\,\Gamma(2/a + 2/b + 1)}$$

3.8 In Forslund's geometric tree volume model, a sample of stems of *Populus tremuloides* (American aspen) gave $\bar{K} = 0.3$. Find the value of a (assuming $b = 1$) and then the volume equation corresponding to (3.60).

3.9 Find the volume under the surface $z = \sin(x + y)$, between the limits $x = -\tfrac{1}{2}\pi$, $x = \tfrac{1}{2}\pi$, $y = 0$, $y = \tfrac{1}{2}\pi$ (see Chapter 4 for the method of integrating this function).

3.10 Evaluate

$$\int_0^a \int_0^{\sqrt{(a^2 - y^2)}} (x^2 + y^2) \, dx \, dy$$

4

Trigonometric and related functions

Trigonometric functions and their biological uses

The trigonometric functions of x are sin x, cos x, tan x, cosec x, sec x, cot x. In words, these functions are: **sine, cosine, tangent, cosecant, secant, cotangent,** of x. You will have already met at least the first three of these in relation to an acute angle, θ, of a right-angled triangle. In Fig. 4.1 is shown a right-angled triangle ABC with the right-angle at C. In relation to the acute angle A, the three sides of the triangle are named thus:

side BC is called 'opposite'
side AC is called 'adjacent'
side AB is called 'hypotenuse'

Then

$$\sin \theta = \frac{\text{opposite}}{\text{hypotenuse}}, \qquad \cos \theta = \frac{\text{adjacent}}{\text{hypotenuse}}, \qquad \tan \theta = \frac{\text{opposite}}{\text{adjacent}}$$

$$\operatorname{cosec} \theta = \frac{\text{hypotenuse}}{\text{opposite}}, \qquad \sec \theta = \frac{\text{hypotenuse}}{\text{adjacent}}, \qquad \cot \theta = \frac{\text{adjacent}}{\text{opposite}}$$

Thus it is seen that the last three functions are merely the reciprocals of the first three, i.e.

$$\operatorname{cosec} \theta = \frac{1}{\sin \theta} \qquad \sec \theta = \frac{1}{\cos \theta} \qquad \cot \theta = \frac{1}{\tan \theta}$$

Also it is easy to verify that

$$\tan \theta = \frac{\sin \theta}{\cos \theta} \qquad \cot \theta = \frac{\cos \theta}{\sin \theta}$$

Defining these six trigonometric functions in relation to a right-angled triangle restricts consideration of them to values of θ in the range $0° \leqslant \theta \leqslant 90°$. There is no reason, however, to prevent us from considering

116

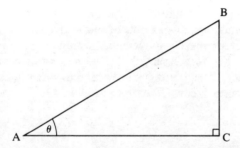

Figure 4.1 A right-angled triangle with acute angle θ.

a trigonometric function of θ when θ exceeds 90°, or even where θ is a 'negative' angle; and the second section of this chapter will be devoted to such general angles and their trigonometric functions.

For the moment, however, let us assume that we can plot a graph of sin θ against θ for the range $0° \leqslant \theta \leqslant 720°$; the curve is shown in Figure 4.2. We see that the portion of the curve from 0° to 360° is repeated exactly from 360° to 720°, and indeed the curve will continue to be repeated every 360°. The curve closely resembles a waveform; the wavelength is 360°, and the amplitude is 2 (from -1 to $+1$). This curve is so well known that any curve of this type is referred to as a 'sine curve', and probably its most important property is the exact repetition every 360°.

All trigonometric functions are cyclic in this way, and this property, together with the fact that they are associated with angular measure, makes for their practical importance. In biology, trigonometric functions arise when considering the radiation environment of plants and animals; days and nights are a cyclic phenomenon and the intensity of radiation on a surface involves the use of angular measure. Population dynamics, which can only be satisfactorily studied by the use of mathematics, often needs to

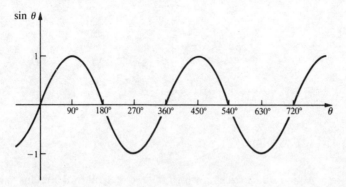

Figure 4.2 Curve of the function $f(\theta) = \sin \theta$ (θ in degrees).

117

explain why some populations oscillate, and this naturally leads to trigonometric functions; an example of this is given in Chapter 8, page 269. Again, the study of muscle movement in animals involves trigonometric functions. It is therefore essential for the quantitative biologist to study trigonometric functions; but their theory and use are not as straightforward as for those functions considered hitherto.

The general angle

Radian measure

In Figure 4.2 is shown the graph of the function $\sin \theta$ plotted against θ. We could, if we wished, put $y = \sin \theta$ and then the vertical axis would be y, and the graph would be that of the function $y = \sin \theta$. Now when drawing graphs of other functions, e.g. $y = ae^{bx}$, both axes have been a scale of natural numbers. In the present case, however, although the vertical axis is a scale of natural numbers, the horizontal axis is not. The scale of degrees is quite arbitrary, but has historical significance. The basis of degrees as an angular measure is that a complete circle subtends an angle of $360°$ at its centre, and so a right-angle, which is the angle subtended by a quarter of a circle, is $360°/4 = 90°$. The question then immediately arises: 'Is it possible to have a more natural angular measure?'

What needs to be done is to have another scale of angle measurement whose unit is defined by some definite (i.e. non-arbitrary) quantity. Consider a circle of radius r (Fig. 4.3) and centre O; A and B are two points on the circle such that the length of the arc between them is equal to the radius of the circle, r. Then angle AÔB is defined as unity on the natural

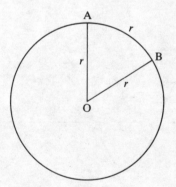

Figure 4.3 A sector of a circle in which the arc is equal to the radius. The angle AÔB is 1 radian.

scale of angular measure, and the unit is known as a radian (rad). Thus

$$A\hat{O}B = 1 \text{ radian}$$

Now we need to know the relationship between degrees and radians. The arc AB is part of the circumference of the circle, is equal to the radius, r, and subtends an angle of 1 rad at the centre. The whole circumference is $2\pi r$ and this subtends an angle of $360°$ at the centre. Hence if

$$r \text{ (of circumference) subtends 1 rad}$$

then

$$2\pi r \text{ (of circumference) subtends } 2\pi \text{ rad}$$

So

$$(2\pi) \text{ rad} = 360°$$

hence

$$1 \text{ rad} = \frac{360°}{2\pi} = \frac{180°}{\pi} \simeq 57°18' \qquad (4.1)$$

and

$$1° = \left(\frac{2\pi}{360}\right) \text{ rad} = \left(\frac{\pi}{180}\right) \text{ rad} \simeq 0.01745 \text{ rad} \qquad (4.2)$$

Since $360° = (2\pi)$ rad, we have $180° = \pi$ rad, $90° = (\pi/2)$ rad, $45° = (\pi/4)$ rad, $270° = (3\pi/2)$ rad, and so on. In Figure 4.2, if the horizontal axis is measured in radians instead of degrees, then the sine curve will appear as in Figure 4.4. Now, both axes may be regarded as natural numbers; notice that we have relabelled the horizontal axis x rather than θ. Although in

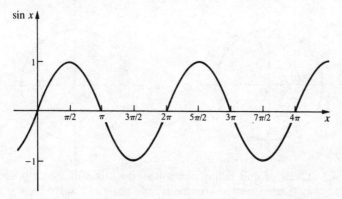

Figure 4.4 Curve of the function $f(x) = \sin x$ (x in radians).

119

many cases it does not matter whether angles are measured in degrees or radians, it is preferable to use radians when in doubt; and it is conventional to designate a scale of angles in a small roman letter when measured in radians, and a Greek letter when the angle is measured in degrees.

The general angle and its trigonometric functions

Consider a circle whose centre is placed at the origin of $x - y$ axes, and whose radius is unity (Fig. 4.5). Let there also be a point P on the circumference, and let the straight line OP subtend an angle θ with the x-axis, measured in an *anti clockwise* direction.

CONDITION $0° < \theta < 90°(0 \text{ RAD} < x < \pi/2 \text{ RAD})$

Now if θ is acute, point P is in the upper right-hand quadrant of the co-ordinate axes, as shown in Figure 4.5a. The triangle OPQ is right-angled at Q and so from previous definitions (page 116)

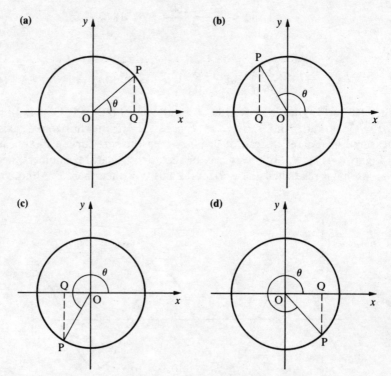

Figure 4.5 Circle of unit radius, centred on the origin of the co-ordinate axes defining the trigonometrical functions of angle θ: (a) $0° < \theta < 90°$, (b) $90° < \theta < 180°$, (c) $180° < \theta < 270°$, (d) $270° < \theta < 360°$.

$$\sin\theta = \frac{PQ}{OP} = \frac{PQ}{1} = PQ \quad \text{(a measurement on the } y\text{-axis)}$$

$$\cos\theta = \frac{OQ}{OP} = \frac{OQ}{1} = OQ \quad \text{(a measurement of the } x\text{-axis)}$$

$$\tan\theta = \frac{PQ}{OQ} \qquad \text{(ratio of measurement on the } y\text{-axis}$$
$$\text{to the measurement on the } x\text{-axis)}$$

Thus the definitions of the trigonometrical functions can be restated in terms of a circle whose centre is at the origin of the usual co-ordinate axes, and in this way we may define these functions for angles in excess of $90°$, $\pi/2$ rad. In the following, we shall only deal with the sine, cosine, and tangent. The remaining three functions may be obtained from these quite easily as reciprocals (page 116).

Before leaving the upper right-hand quadrant, we note that

$$\begin{array}{llll}
\sin 0° = 0 & (PQ = 0) & \sin 90° = 1 & (PQ = 1) \\
\cos 0° = 1 & (OQ = 1) & \cos 90° = 0 & (OQ = 0) \\
\tan 0° = 0 & (PQ/OQ = 0) & \tan 90° \text{ is infinitely large}
\end{array}$$

These results can be checked on a scientific calculator or in the usual tables of trigonometric functions, and similarly we may find the values of sine, cosine, and tangent of any angle lying in the range $0° < \theta < 90°$.

CONDITION $90° < \theta < 180°$ ($\pi/2$ RAD $< x < \pi$ RAD)

$\sin\theta$: Since PQ is positive (Fig. 4.5b), $\sin\theta$ is positive in this range. The length of OP is *always* positive. Now $\sin 90° = 1$ (PQ = 1), and $\sin 180° = 0$ (PQ = 0). So

$$\sin\theta = \sin(180 - \theta)$$

$\cos\theta$: Since OQ is negative, $\cos\theta$ is negative, in this range. Now $\cos 90° = 0$ (OQ = 0), and $\cos 180° = -1$ (OQ = -1). So

$$\cos\theta = -\cos(180 - \theta)$$

$\tan\theta$:

$$\tan\theta = \frac{\sin\theta}{\cos\theta} = \frac{\sin(180 - \theta)}{-\cos(180 - \theta)}$$

So

$$\tan\theta = -\tan(180 - \theta)$$

CONDITION $180° < \theta < 270°$ (π RAD $< x < 3\pi/2$ RAD)

$\sin\theta$: Since PQ is negative (Fig. 4.5c), $\sin\theta$ is negative in this range. Now

121

$\sin 180° = 0$ ($PQ = 0$), and $\sin 270° = -1$ ($PQ = -1$). So

$$\sin \theta = -\sin(\theta - 180)$$

$\cos \theta$: Since OQ is negative, $\cos \theta$ is negative in this range. Now $\cos 180° = -1$ ($OQ = -1$), and $\cos 270° = 0$ ($OQ = 0$). So

$$\cos \theta = -\cos(\theta - 180)$$

$\tan \theta$:

$$\tan \theta = \frac{\sin \theta}{\cos \theta} = \frac{-\sin(\theta - 180)}{-\cos(\theta - 180)}$$

So

$$\tan \theta = \tan(\theta - 180)$$

CONDITION $270° < \theta < 360°$ ($3\pi/2$ RAD $< x < 2\pi$ RAD)

$\sin \theta$: Since PQ is negative (Fig. 4.5d), $\sin \theta$ is negative in this range. Now $\sin 270° = -1$ ($PQ = -1$), and $\sin 360° = 0$ ($PQ = 0$). So

$$\sin \theta = -\sin(360 - \theta)$$

$\cos \theta$: Since OQ is positive, $\cos \theta$ is positive in this range. Now $\cos 270° = 0$ ($OQ = 0$), and $\cos 360° = 1$ ($OQ = 1$). So

$$\cos \theta = \cos(360 - \theta)$$

$\tan \theta$:

$$\tan \theta = \frac{\sin \theta}{\cos \theta} = \frac{-\sin(360 - \theta)}{\cos(360 - \theta)}$$

So

$$\tan \theta = -\tan(360 - \theta)$$

CONDITION $\theta > 360°$ ($x > 2\pi$ RAD)

If $\theta > 360°$ this means that point P has completed at least one revolution around the circle in an anticlockwise direction. If $360° < \theta \leqslant 720°$, the trigonometric functions of θ can be found by first deducting $360°$ from θ, and then by using the above rules. So

$$\sin \theta = \sin(\theta - 360)$$

$$\cos \theta = \cos(\theta - 360)$$

$$\tan \theta = \tan(\theta - 360)$$

If $720° < \theta \leqslant 1080°$, then, e.g. $\sin \theta = \sin(\theta - 720)$, and so on. It is thus evident that trigonometric functions repeat themselves every $360°$, or, more important, every 2π radians.

NEGATIVE ANGLES

These are obtained by rotating P in a clockwise direction. So if $0° > \theta \geqslant -360°$, we have, for example

$$\sin \theta = \sin(\theta + 360)$$

If $-360° > \theta \geqslant -720°$, we have, for example

$$\tan \theta = \tan(\theta + 720)$$

and so on.

It is now apparent that the trigonometric functions are basically defined in relation to a cricle, and that their definition in connexion with a triangle is a special case. Because of their association with a circle, trigonometric functions are sometimes called **circular functions**.

Graphs of all six trigonometric functions are shown in Figure 4.6.

Trigonometric identities

Before dealing with the identities in this section, it is necessary to consider an item of notation. If, say, $\sin x$ is raised to the power n, that is $(\sin x)^n$, it is customary to write it as $\sin^n x$. Thus

$$\sin^2 x = (\sin x)^2 = (\sin x)(\sin x)$$

Distinguish carefully between $\sin^2 x$ and $\sin x^2$ the latter is

$$\sin x^2 = \sin(x^2) = \sin(xx)$$

Identities involving a single angle

In relation to the circle in Figure 4.5, we have shown that $\sin \theta = PQ$ and $\cos \theta = OQ$ (page 121). Now in the right-angled triangle OPQ, by Pythagoras's theorem

$$OQ^2 + PQ^2 = OP^2$$

But $OQ = \cos \theta$, $PQ = \sin \theta$, and $OP = 1$ (radius of the circle). Hence

$$\cos^2 \theta + \sin^2 \theta = 1 \tag{4.3}$$

This is the **fundamental identity** in trigonometry; and two other useful ones can be derived from it. Divide both sides of (4.3) by $\cos^2 \theta$.

$$1 + \frac{\sin^2 \theta}{\cos^2 \theta} = \frac{1}{\cos^2 \theta}$$

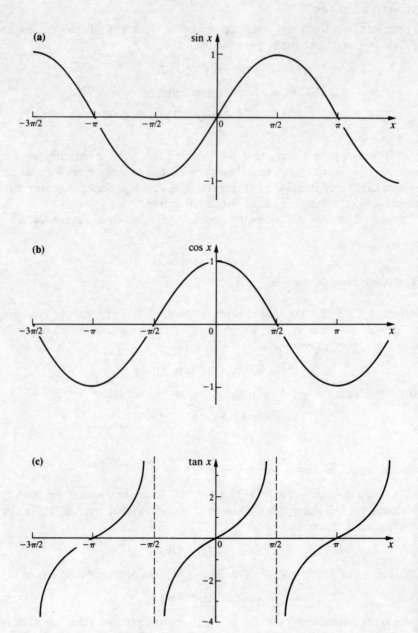

Figure 4.6 The trigonometrical function of x over the range $-3\pi/2 < x < 3\pi/2$: (a) $\sin x$, (b) $\cos x$, (c) $\tan x$, (d) $\csc x$, (e) $\sec x$, (f) $\cot x$.

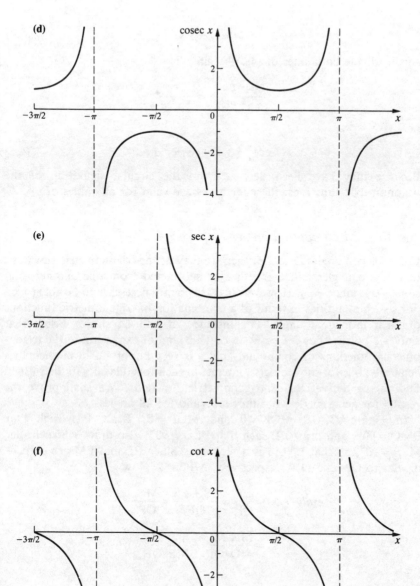

(d)

cosec x

$-3\pi/2$ $\quad$ $-\pi$ $\quad$ $-\pi/2$ $\quad$ 0 $\quad$ $\pi/2$ $\quad$ π $\quad$ x

(e)

sec x

$-3\pi/2$ $\quad$ $-\pi$ $\quad$ $-\pi/2$ $\quad$ 0 $\quad$ $\pi/2$ $\quad$ π $\quad$ x

(f)

cot x

$-3\pi/2$ $\quad$ $-\pi$ $\quad$ $-\pi/2$ $\quad$ 0 $\quad$ $\pi/2$ $\quad$ π $\quad$ x

i.e.

$$1 + \tan^2 \theta = \sec^2 \theta \qquad (4.4)$$

Again, divide both sides of (4.3) by $\sin^2 \theta$:

$$\frac{\cos^2 \theta}{\sin^2 \theta} + 1 = \frac{1}{\sin^2 \theta}$$

i.e.

$$\cot^2 \theta + 1 = \operatorname{cosec}^2 \theta \qquad (4.5)$$

Because they have been derived from the circular definition of the trigonometric functions, these identities are valid for all values of θ.

Identities of compound angles

Consider two angles, θ and ϕ; then we may define trigonometric functions of the two angles added together or subtracted from one another, e.g. $\sin(\theta + \phi)$, $\tan(\theta - \phi)$. However, it is frequently necessary to be able to express such quantities in terms of a combination of trigonometric functions of each individual angle. For instance, it will be shown below that $\sin(\theta + \phi) = \sin \theta \cos \phi + \cos \theta \sin \phi$. This kind of expansion of the trigonometric functions of compound angles is very important in mathematics generally, but for the biologist, interest in such formulae lies in their use in finding the derivatives of trigonometric functions. We shall prove the results for acute angles, but they are valid for all angles.

In Figure 4.7, let $M\hat{O}N = \theta$ and $M\hat{O}P = \phi$. Draw PM such that $P\hat{M}O = 90°$, and draw PL such that $P\hat{L}O = 90°$; also draw MS such that $M\hat{S}P = 90°$, so that LNMS is a rectangle. Since PL and PM are perpendicular to ON and OM, respectively, $M\hat{P}S = \theta$. Now

$$\sin(\theta + \phi) = \frac{LP}{OP} = \frac{LS + PS}{OP} = \frac{MN}{OP} + \frac{PS}{OP}$$

$$= \frac{MN}{MO} \frac{MO}{OP} + \frac{PS}{MP} \frac{MP}{OP}$$

i.e.

$$\sin(\theta + \phi) = \sin \theta \cos \phi + \cos \theta \sin \phi \qquad (4.6)$$

$$\cos(\theta + \phi) = \frac{LO}{OP} = \frac{NO - LN}{OP} = \frac{NO}{OP} - \frac{MS}{OP}$$

$$= \frac{NO}{MO} \frac{MO}{OP} - \frac{MS}{MP} \frac{MP}{OP}$$

126

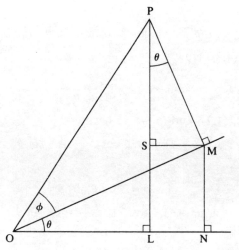

Figure 4.7 Diagram to derive the trigonometric functions of the compound angle $(\theta \pm \phi)$.

i.e.

$$\cos(\theta + \phi) = \cos\theta \cos\phi - \sin\theta \sin\phi \qquad (4.7)$$

$$\tan(\theta + \phi) = \frac{\sin(\theta + \phi)}{\cos(\theta + \phi)} = \frac{\sin\theta \cos\phi + \cos\theta \sin\phi}{\cos\theta \cos\phi - \sin\theta \sin\phi}$$

Divide numerator and denomintor of the right-hand side by $\cos\theta \cos\phi$; then we have

$$\tan(\theta + \phi) = \frac{\dfrac{\sin\theta \cos\phi}{\cos\theta \cos\phi} + \dfrac{\cos\theta \sin\phi}{\cos\theta \cos\phi}}{\dfrac{\cos\theta \cos\phi}{\cos\theta \cos\phi} - \dfrac{\sin\theta \sin\phi}{\cos\theta \cos\phi}}$$

$$\tan(\theta + \phi) = \frac{\tan\theta + \tan\phi}{1 - \tan\theta \tan\phi} \qquad (4.8)$$

Now put $\phi = -\phi$ in (4.6) then

$$\sin\{\theta + (-\phi)\} = \sin\theta \cos(-\phi) + \cos\theta \sin(-\phi)$$

But $\theta + (-\phi) = \theta - \phi$, and from the definitions of the general angle given earlier (page 120), $\cos(-\phi) = \cos\phi$, $\sin(-\phi) = -\sin\phi$. So

$$\sin(\theta - \phi) = \sin\theta \cos\phi - \cos\theta \sin\phi \qquad (4.9)$$

Put $\phi = -\phi$ in (4.7), then

$$\cos\{\theta + (-\phi)\} = \cos\theta \cos(-\phi) - \sin\theta \sin(-\phi)$$

so

$$\cos(\theta - \phi) = \cos \theta \cos \phi + \sin \theta \sin \phi \qquad (4.10)$$

Put $\phi = -\phi$ in (4.8). Now $\tan(-\phi) = -\tan \phi$. So

$$\tan(\theta - \phi) = \frac{\tan \theta - \tan \phi}{1 + \tan \theta \tan \phi} \qquad (4.11)$$

If $\phi = \theta$ in (4.6), (4.7), and (4.8), we have

$$\sin(\theta + \theta) = \sin \theta \cos \theta + \cos \theta \sin \theta$$

i.e.

$$\sin 2\theta = 2 \sin \theta \cos \theta \qquad (4.12)$$

$$\cos(\theta + \theta) = \cos \theta \cos \theta - \sin \theta \sin \theta$$

i.e.

$$\cos 2\theta = \cos^2 \theta - \sin^2 \theta \qquad (4.13)$$

$$\tan(\theta + \theta) = \frac{\tan \theta + \tan \theta}{1 - \tan \theta \tan \theta}$$

i.e.

$$\tan 2\theta = \frac{2 \tan \theta}{1 - \tan^2 \theta} \qquad (4.14)$$

Finally, we develop four other results which will be useful later. Adding together (4.6) and (4.9) gives

$$\sin(\theta + \phi) = \sin \theta \cos \phi + \cos \theta \sin \phi$$

$$\text{plus } \sin(\theta - \phi) = \sin \theta \cos \phi - \cos \theta \sin \phi$$

so

$$\sin(\theta + \phi) + \sin(\theta - \phi) = 2 \sin \theta \cos \phi \qquad (4.15)$$

Equation (4.6) minus equation (4.9) gives

$$\sin(\theta + \phi) = \sin \theta \cos \phi + \cos \theta \sin \phi$$

$$\text{minus } \sin(\theta - \phi) = \sin \theta \cos \phi - \cos \theta \sin \phi$$

so

$$\sin(\theta + \phi) - \sin(\theta - \phi) = 2 \cos \theta \sin \phi \qquad (4.16)$$

Adding (4.7) and (4.10) gives

$$\cos(\theta + \phi) = \cos \theta \cos \phi - \sin \theta \sin \phi$$

$$\text{plus } \cos(\theta - \phi) = \cos \theta \cos \phi + \sin \theta \sin \phi$$

so

$$\cos(\theta + \phi) + \cos(\theta - \phi) = 2 \cos \theta \cos \phi \qquad (4.17)$$

Equation (4.7) minus equation (4.10) gives

$$\cos(\theta + \phi) = \cos \theta \cos \phi - \sin \theta \sin \phi$$

$$\text{minus } \cos(\theta - \phi) = \cos \theta \cos \phi + \sin \theta \sin \phi$$

so

$$\cos(\theta + \phi) - \cos(\theta - \phi) = -2 \sin \theta \sin \phi \qquad (4.18)$$

The calculus of trigonometric functions

Limit theorems

Before the trigonometric functions can be differentiated, special trigonometric limits need to be presented. Only the first of these is essential for our purpose, but two allied ones are given for the sake of completeness.

All angles involved in limits, and in the calculus, of trigonometric functions are measured in radians.

THEOREM 4.1 $$\lim_{x \to 0} \left(\frac{\sin x}{x} \right) = 1$$

where the angle x is measured in radians.

Proof. In Figure 4.8, AB is an arc of a circle, centre O and radius r, subtending an angle $x \, (< \pi/2 \text{ rad})$ at O. The line BT is a tangent to the circle at B, extended to meet OA, produced, at T. The line BN is drawn such that BNO is a right angle. Now inspection of Figure 4.8 shows that

$$BN < AB < BT$$

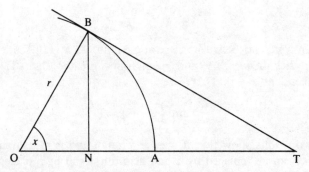

Figure 4.8 Diagram to investigate the limit of $(\sin x)/x$ as x approaches 0.

129

Divide throughout by BO:

$$\frac{BN}{BO} < \frac{AB}{BO} < \frac{BT}{BO}$$

i.e.

$$\sin x < x < \tan x \qquad (4.19)$$

(since x is in radians).

Invert throughout and replace tan x by (sin x)(cos x)

$$\frac{1}{\sin x} > \frac{1}{x} > \frac{\cos x}{\sin x}$$

and multiplying through by sin x gives

$$1 > \frac{\sin x}{x} > \cos x$$

Thus $(\sin x)/x$ lies between 1 and cos x. But as $x \to 0$, cos $x \to 1$; hence $(\sin x)/x$ lies between 1 and a quantity less than 1, but which approaches 1 as $x \to 0$. Therefore

$$\lim_{x \to 0} \left(\frac{\sin x}{x}\right) = 1 \qquad (4.20)$$

THEOREM 4.2 $\qquad\qquad \lim_{x \to 0} \left(\frac{\tan x}{x}\right) = 1$

where the angle x is measured in radians.

Proof. Divide through the inequality (4.19) by tan x:

$$\cos x < \frac{x}{\tan x} < 1$$

Invert throughout:

$$\sec x > \frac{\tan x}{x} > 1$$

Thus $(\tan x)/x$ lies between sec x and 1. But as $x \to 0$, sec $x \to 1$; hence $(\tan x)/x$ lies between 1 and a quantity greater than 1, but which approaches 1 as $x \to 0$. Therefore

$$\lim_{x \to 0} \left(\frac{\tan x}{x}\right) = 1 \qquad (4.21)$$

These two results show that if x is very small and is measured in radians then sin x can be replaced by x and also tan x can be replaced by x. The smaller the value of x, the more accurate will be the results. There is a fair measure of agreement between x, sin x, and tan x for angles up to about 0.087 27 radians (5°).

THEOREM 4.3 $\qquad \lim_{x \to 0} \left(\frac{1 - \cos x}{x^2} \right) = \tfrac{1}{2}$

where the angle x is measured in radians.

Proof. Replace θ by $\tfrac{1}{2}x$ in identity (4.13), then we have

$$\cos x = \cos^2 \tfrac{1}{2}x - \sin^2 \tfrac{1}{2}x \qquad (4.22)$$

From identity (4.3), we have

$$\cos^2 \tfrac{1}{2}x = 1 - \sin^2 \tfrac{1}{2}x$$

So substituting for $\cos^2 \tfrac{1}{2}x$ in (4.22) yields

$$\cos x = 1 - \sin^2 \tfrac{1}{2}x - \sin^2 \tfrac{1}{2}x$$

i.e.

$$1 - \cos x = 2 \sin^2 \tfrac{1}{2}x$$

Hence

$$\frac{1 - \cos x}{x^2} = \frac{2 \sin^2 \tfrac{1}{2}x}{x^2} = 2 \left(\frac{\sin \tfrac{1}{2}x}{x} \right)^2$$

and so

$$\frac{1 - \cos x}{x^2} = \tfrac{1}{2} \left(\frac{\sin \tfrac{1}{2}x}{\tfrac{1}{2}x} \right)^2$$

Since

$$\lim_{x \to 0} \left(\frac{\sin \tfrac{1}{2}x}{\tfrac{1}{2}x} \right) = 1$$

then

$$\lim_{x \to 0} \left(\frac{1 - \cos x}{x^2} \right) = \tfrac{1}{2}(1)^2 = \tfrac{1}{2} \qquad (4.23)$$

Differentiation

We shall differentiate sin x and cos x from first principles, and then use these results combined with the usual rules to differentiate the remaining trigonometric functions.

131

Example 4.1   Find

$$\text{(a)} \ \frac{d(\sin \ x)}{dx} \qquad \text{(b)} \ \frac{d(\cos \ x)}{dx} \qquad \text{(c)} \ \frac{d(\tan \ x)}{dx}$$

$$\text{(d)} \ \frac{d(\cosec \ x)}{dx} \qquad \text{(e)} \ \frac{d(\sec \ x)}{dx} \qquad \text{(f)} \ \frac{d(\cot \ x)}{dx} \qquad (4.24)$$

(a)   Put
$$y = \sin \ x \qquad (4.24)$$

Let there be a small increase of δx in x and a corresponding increment of δy in y; then

$$y + \delta y = \sin(x + \delta x) \qquad (4.25)$$

Subtracting (4.24) from (4.25) gives

$$\delta y = \sin(x + \delta x) - \sin \ x \qquad (4.26)$$

Now we use identity (4.16) with $\theta = x + \tfrac{1}{2} \ \delta x$ and $\phi = \tfrac{1}{2} \ \delta x$. This gives

$$\sin\{ (x + \tfrac{1}{2} \ \delta x) + \tfrac{1}{2} \ \delta x\} - \sin\{ (x + \tfrac{1}{2} \ \delta x) - \tfrac{1}{2} \ \delta x\}$$
$$= 2 \ \cos(x + \tfrac{1}{2} \ \delta x) \sin \tfrac{1}{2} \ \delta x$$

i.e.

$$\sin(x + \delta x) - \sin \ x = 2 \ \cos(x + \tfrac{1}{2} \ \delta x) \sin \tfrac{1}{2} \ \delta x$$

Substituting into the right-hand side of (4.26), we have

$$\delta y = 2 \ \cos(x + \tfrac{1}{2} \ \delta x) \sin \tfrac{1}{2} \ \delta x$$

Divide both sides by δx:

$$\frac{\delta y}{\delta x} = \frac{2 \ \cos(x + \tfrac{1}{2} \ \delta x) \sin \tfrac{1}{2} \ \delta x}{\delta x}$$

Divide both numerator and denominator of the right-hand side by 2:

$$\frac{\delta y}{\delta x} = \cos(x + \tfrac{1}{2} \ \delta x) \frac{\sin \tfrac{1}{2} \ \delta x}{\tfrac{1}{2} \ \delta x}$$

So

$$\frac{dy}{dx} = \lim_{\delta x \to 0} \left\{ \cos(x + \tfrac{1}{2} \ \delta x) \frac{\sin \tfrac{1}{2} \ \delta x}{\tfrac{1}{2} \ \delta x} \right\} = \cos \ x$$

since

$$\lim_{\delta x \to 0} \left(\frac{\sin \tfrac{1}{2} \ \delta x}{\tfrac{1}{2} \ \delta x} \right) = 1$$

(page 129). Thus

$$\frac{d(\sin \ x)}{dx} = \cos \ x \qquad (4.27$$

132

(b) Put
$$y = \cos x. \tag{4.28}$$

Let there be a small increase of δx in x and a corresponding increment of δy in y; then

$$y + \delta y = \cos(x + \delta x) \tag{4.29}$$

Subtracting (4.28) and (4.29):

$$\delta y = \cos(x + \delta x) - \cos x$$

Beginning with the identity (4.18), we obtain in a similar manner to the previous example

$$\delta y = -2 \sin(x + \tfrac{1}{2} \delta x) \sin \tfrac{1}{2} \delta x$$

Dividing both sides by δx, and then the right-hand side by 2:

$$\frac{\delta y}{\delta x} = -\sin(x + \tfrac{1}{2} \delta x) \frac{\sin \tfrac{1}{2} \delta x}{\tfrac{1}{2} \delta x}$$

$$\frac{dy}{dx} = \lim_{\delta x \to 0} \left\{ -\sin(x + \tfrac{1}{2} \delta x) \frac{\sin \tfrac{1}{2} \delta x}{\tfrac{1}{2} \delta x} \right\} = -\sin x$$

Thus

$$\frac{d(\cos x)}{dx} = -\sin x \tag{4.30}$$

(c)
$$\tan x = \frac{\sin x}{\cos x}$$

Hence, applying the quotient rule, we have

$$\frac{d(\tan x)}{dx} = \frac{(\cos x)\, d(\sin x)/dx - (\sin x)\, d(\cos x)/dx}{\cos^2 x}$$

Therefore

$$\frac{d(\tan x)}{dx} = \frac{\cos^2 x + \sin^2 x}{\cos^2 x} = \frac{1}{\cos^2 x}$$

(using the fundamental identity, (4.3), and so

$$\frac{d(\tan x)}{dx} = \sec^2 x \tag{4.31}$$

(d)
$$\operatorname{cosec} x = \frac{1}{\sin x}$$

Applying the quotient rule, we have

$$\frac{d(\operatorname{cosec} x)}{dx} = \frac{-\cos x}{\sin^2 x} = -\frac{1}{\sin x} \frac{\cos x}{\sin x}$$

133

Thus

$$\frac{d(\text{cosec } x)}{dx} = -\text{cosec } x \cot x \qquad (4.32)$$

(e)
$$\sec x = \frac{1}{\cos x}$$

Again, one could apply the quotient rule, but for a change this example will be worked by the function of a function rule. Put $y = \sec x$, i.e. $y = 1/\cos x$, and put $u = \cos x$, so that

$$y = \frac{1}{u}$$

Then

$$\frac{du}{dx} = -\sin x \quad \text{and} \quad \frac{dy}{du} = -\frac{1}{u^2}$$

Thus

$$\frac{dy}{dx} = \left(-\frac{1}{u^2}\right)\left(-\sin x\right) = \frac{\sin x}{\cos^2 x} = \frac{1}{\cos x}\frac{\sin x}{\cos x}$$

and so

$$\frac{d(\sec x)}{dx} = \sec x \tan x \qquad (4.33)$$

(f)
$$\cot x = \frac{\cos x}{\sin x}$$

Applying the quotient rule:

$$\frac{d(\cot x)}{dx} = \frac{(\sin x)\, d(\cos x)/dx - (\cos x)\, d(\sin x)/dx}{\sin^2 x}$$

$$= \frac{-\sin^2 x - \cos^2 x}{\sin^2 x} = -\frac{(\sin^2 x + \cos^2 x)}{\sin^2 x}$$

So, using the fundamental identity, (4.3), we finally have

$$\frac{d(\cot x)}{dx} = -\text{cosec}^2 x \qquad (4.34)$$

More complex trigonometric functions can be differentiated by the usual rules applied to the above basic results.

Example 4.2 Differentiate with respect to x:

(a) $\sin^3 x$ *(b)* $\cos x^4$ *(c)* $\sec(a + bx)$

(a)   Put $y = \sin^3 x$ and $u = \sin x$, then $y = u^3$; so

$$\frac{du}{dx} = \cos x$$

and

$$\frac{dy}{du} = 3u^2$$

Hence

$$\frac{dy}{dx} = 3u^2 \cos x$$

and so

$$\frac{d(\sin^3 x)}{dx} = 3 \sin^2 x \cos x$$

(b)   Put $y = \cos x^4$ and $u = x^4$, then $y = \cos u$; so

$$\frac{du}{dx} = 4x^3 \quad \text{and} \quad \frac{dy}{du} = -\sin u$$

Hence

$$\frac{dy}{dx} = (-\sin u)4x^3$$

and so

$$\frac{d(\cos x^4)}{dx} = -4x^3 \sin x^4$$

(c)   Put $y = \sec(a + bx)$ and $u = a + bx$, then $y = \sec u$; so

$$\frac{du}{dx} = b$$

and

$$\frac{dy}{du} = \sec u \tan u$$

$$\frac{dy}{dx} = b \sec u \tan u$$

i.e.

$$\frac{d\{\sec(a + bx)\}}{dx} = b \sec(a + bx) \tan(a + bx)$$

135

Integration

Reversal of the equations (4.27), and (4.30) to (4.34) inclusive, implies that the integrals of the right-hand sides of these equations are equal to the left-hand sides. In particular, we have

$$\int (\cos x)\, dx = \sin x + c \tag{4.35}$$

and

$$\int (\sin x)\, dx = -\cos x + c \tag{4.36}$$

The two results are fundamental to the integration of trigonometric functions, and integrals of the remaining functions are found by applying various methods to the two standard integrals (4.35) and (4.36).

Example 4.3 Find

(a) $\int (\tan x)\, dx$ (b) $\int (\operatorname{cosec} x)\, dx$ (c) $\int (\sec x)\, dx$ (d) $\int (\cot x)\, dx$

(a) $$\int (\tan x)\, dx = \int \frac{\sin x}{\cos x}\, dx$$

Make the substitution $u = \cos x$, then

$$\frac{du}{dx} = -\sin x$$

and so

$$\sin x = -\frac{du}{dx}$$

Therefore

$$\int \frac{(\sin x)\, dx}{\cos x} = \int \frac{-(du/dx)\, dx}{u} = -\int \frac{du}{u} = -\log_e u + c$$
$$= -\log_e (\cos x) + c$$

Now

$$-\log_e (\cos x) = \log_e (1/\cos x)$$

so

$$\int (\tan x)\, dx = \log_e (\sec x) + c \tag{4.37}$$

136

(b)
$$\int (\operatorname{cosec} x)\, dx = \int \frac{dx}{\sin x}$$

Now from (4.12)

$$\sin x = 2 \sin \tfrac{1}{2} x \cos \tfrac{1}{2} x$$

so

$$\int \frac{dx}{\sin x} = \tfrac{1}{2} \int \frac{dx}{\sin \tfrac{1}{2} x \cos \tfrac{1}{2} x}$$

Divide numerator and denominator of the right-hand side by $\cos^2 \tfrac{1}{2} x$; then

$$\tfrac{1}{2} \int \frac{(\cos^2 \tfrac{1}{2} x)\, dx}{(\sin \tfrac{1}{2} x)(\cos \tfrac{1}{2} x)^{-1}(\cos \tfrac{1}{2} x)(\cos \tfrac{1}{2} x)^{-1}} = \tfrac{1}{2} \int \frac{\sec^2 \tfrac{1}{2} x}{\tan \tfrac{1}{2} x}\, dx$$

Put $u = \tan \tfrac{1}{2} x$, then

$$\frac{du}{dx} = \tfrac{1}{2} \sec^2 \tfrac{1}{2} x$$

i.e.

$$dx = \frac{2\, du}{\sec^2 \tfrac{1}{2} x}$$

Hence

$$\tfrac{1}{2} \int \frac{(\sec^2 \tfrac{1}{2} x)\, dx}{\tan \tfrac{1}{2} x} = \tfrac{1}{2} \int \frac{\sec^2 \tfrac{1}{2} x 2 (\sec^2 \tfrac{1}{2} x)^{-1}\, du}{u} = \int \frac{du}{u} = \log_e u + c$$

and so

$$\int (\operatorname{cosec} x)\, dx = \log_e (\tan \tfrac{1}{2} x) + c \qquad (4.38)$$

(c) Multiply both numerator and denominator by $(\sec x + \tan x)$, then

$$\int (\sec x)\, dx = \int \frac{\sec x (\sec x + \tan x)}{\sec x + \tan x}\, dx \qquad (4.39)$$

Now make substitution $u = \sec x + \tan x$, then

$$\frac{du}{dx} = \sec x \tan x + \sec^2 x$$

$$= \sec x(\sec x + \tan x)$$

$$= u \sec x$$

Hence

$$dx = \frac{du}{u \sec x}$$

137

Substituting into the right-hand side of (4.39) gives

$$\int (\sec\ x)\ \mathrm{d}x = \int \frac{(\sec\ x)u(u\ \sec\ x)^{-1}\ \mathrm{d}u}{u} = \int \frac{\mathrm{d}u}{u} = \log_e u + c$$

and so

$$\int (\sec\ x)\ \mathrm{d}x = \log_e (\sec\ x + \tan\ x) + c \qquad (4.40)$$

(d)
$$\int (\cot\ x)\ \mathrm{d}x = \int \frac{\cos\ x}{\sin\ x}\ \mathrm{d}x$$

Make the substitution $u = \sin\ x$, then

$$\frac{\mathrm{d}u}{\mathrm{d}x} = \cos\ x$$

and so

$$\int \frac{(\cos\ x)\ \mathrm{d}x}{\sin\ x} = \int \frac{(\mathrm{d}u/\mathrm{d}x)\ \mathrm{d}x}{u} = \int \frac{\mathrm{d}u}{u} = \log_e u + c$$

Therefore

$$\int (\cot\ x)\ \mathrm{d}x = \log_e (\sin\ x) + c \qquad (4.41)$$

Inverse trigonometric functions

Inverse functions

When y is expressed as a function of x, $y = f(x)$, a useful idea is to find the **inverse** of f, often written as f^{-1}. It is the function that has the opposite effect to f; for example, if $f(x) = x^3$ then f^{-1} will take the cube root of x, i.e. $f^{-1}(x) = x^{1/3}$. Carrying out the operation f followed by f^{-1} gets us back just where we started (as does f^{-1} followed by f). In our example, if we begin with a value x, then $y = f(x) = x^3$; now take the inverse of x^3, which is $f^{-1}(y) = f^{-1}(x^3)$ and so is equal to $(x^3)^{1/3}$ or x.

Essentially the inverse of $y = f(x)$ is $x = f(y)$, with the variables changed round; the inverse of $y = x^3$ could also be written $y^{1/3} = x$. One important class of inverses involves the trigonometric functions, and these provide very useful basic forms for differentiation and standard integrals.

The inverse sine, or arcsin

The inverse of $y = \sin\ x$ is written

$$y = \sin^{-1}x \qquad (4.42)$$

that is, y is the angle (in radians) whose sine is x, so that we also have $\sin y = x$. Another notation for the inverse, often used instead of $\sin^{-1}$, is

$$y = \arcsin x \qquad (4.43)$$

The graph of $\sin^{-1} x$ is shown in Figure 4.9a; on comparing it with the graph of $\sin x$ (Fig. 4.6a) the 'inverse' relationship between the two functions is obvious.

There are two properties of the arcsin function which we have not met in previous functions studied. Firstly, x lies only in the range $-1 \leqslant x \leqslant 1$; we say that the **domain of the function** lies in this range, and for any value of x outside this range the arcsin function is undefined. In nearly all other functions we have looked at the domain has been $-\infty < x < \infty$, except for the rectangular hyperbola $y = 1/x$ where $x \neq 0$.

Secondly, for any value of x in the domain, there are an infinity of y-values or (arcsin x)-values. Two of these lie between 0 and 2π radians, and each of the others differs from one of these two by a whole multiple of $\pm 2\pi$ radians. Among all these values of $\sin^{-1} x$ there is always one and

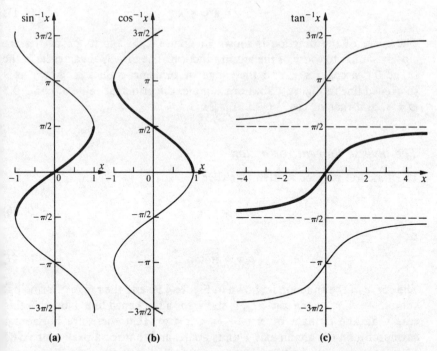

(a)   (b)   (c)

Figure 4.9 The inverse trigonometric functions in the range $-3\pi/2 \leqslant f(x) \leqslant 3\pi/2$: (a) $\sin^{-1} x$, (b) $\cos^{-1} x$, (c) $\tan^{-1} x$. The principal range of each function is shown bold.

139

only one lying between $-\frac{1}{2}\pi$ and $\frac{1}{2}\pi$; this is called the **principal value** of $\sin^{-1}x$ and is numerically its smallest value. For example, when $x = 0.5$, arcsin $x = \pi/6$ and $5\pi/6$, and an infinite number of other values obtainable from these two by adding or subtracting integer multiples of 2π; so

$$y = \pi/6 \pm 2\pi n \quad \text{and} \quad y = 5\pi/6 \pm 2\pi n \qquad n = 0, 1, 2, \ldots$$

The principal value of $\sin^{-1}x$ when $x = 0.5$ is $\pi/6$, and when $x = -0.5$ it is $-\pi/6$. The range of the curve comprising the principal values is shown as a thick line in Figure 4.9a.

The inverse cosine, or arcos

The function may be written as either

$$y = \cos^{-1}x \tag{4.44}$$

or

$$y = \text{arcos } x \tag{4.45}$$

The curve of the function is shown in Figure 4.9b, and its properties are closely similar to those of the arcsin function. The principal values lie in the range $0 \leqslant \text{arcos } x \leqslant \pi$ and the range of principal values is shown as a thickened line in Figure 4.9b. For example, the principal value of $\cos^{-1} 0.5$ is $\pi/3$, and that of $\cos^{-1} (-0.5)$ is $2\pi/3$.

The inverse tangent, or arctan

The function may be written as either

$$y = \tan^{-1}x \tag{4.46}$$

or

$$y = \text{arctan } x \tag{4.47}$$

The curve of the function is shown in Figure 4.9c and the range of principal values, $-\pi/2 < \text{arctan } x < \pi/2$, is shown as a thickened line. Notice in this case that: the domain of x is $-\infty < x < \infty$, that there are horizontal asymptotes on the arctan axis π units apart, and that for a particular value of x there are an infinity of arctan values π units apart.

Functions of $\text{cosec}^{-1}x$, $\sec^{-1}x$, and $\cot^{-1}x$ are also defined but, being of lesser importance, will not be described here.

The calculus of inverse trigonometric functions

Differentiation

Example 4.4   Find

$$(a) \ \frac{d(\sin^{-1}x)}{dx} \qquad (b) \ \frac{d(\cos^{-1}x)}{dx} \qquad (c) \ \frac{d(\tan^{-1}x)}{dx}$$

(a)   Let $y = \sin^{-1}x$; then $x = \sin y$,

$$\frac{dx}{dy} = \cos y \quad \text{or} \quad \frac{dy}{dx} = \frac{1}{\cos y}$$

Now from identity (4.3), $\cos^2 y = 1 - \sin^2 y$, and so $\cos y = \pm\sqrt{(1 - \sin^2 y)}$. Let y have its principal value, which lies in the range $-\tfrac{1}{2}\pi \leqslant y \leqslant \tfrac{1}{2}\pi$. But from Figure 4.6b it is seen that $\cos y$ is positive in this range, and so the positive square root is required. Thus $dy/dx = 1/\sqrt{(1 - \sin^2 y)}$, and because $x = \sin y$ we finally have

$$\frac{d(\sin^{-1} x)}{dx} = \frac{1}{\sqrt{(1 - x^2)}} \qquad (4.48)$$

(b)   Let $y = \cos^{-1} x$; then $x = \cos y$,

$$\frac{dx}{dy} = -\sin y \quad \text{or} \quad \frac{dy}{dx} = -\frac{1}{\sin y}$$

From identity (4.3), $\sin^2 y = 1 - \cos^2 y$, and so $\sin y = \pm\sqrt{(1 - \cos^2 y)}$. Let y have its principal value, which lies in the range $0 \leqslant y \leqslant \pi$. But from Figure 4.6a it is seen that $\sin y$ is positive in this range, and so the positive square root is again required. Thus $dy/dx = -1/\sqrt{(1 - \cos^2 y)}$, and because $x = \cos y$ we finally have

$$\frac{d(\cos^{-1} x)}{dx} = -\frac{1}{\sqrt{(1 - x^2)}} \qquad (4.49)$$

(c)   Let $y = \tan^{-1}x$; then $x = \tan y$,

$$\frac{dx}{dy} = \sec^2 y \quad \text{or} \quad \frac{dy}{dx} = \frac{1}{\sec^2 y}$$

From identity (4.4), $\sec^2 y = 1 + \tan^2 y$, and so, because $x = \tan y$, we have

$$\frac{d(\tan^{-1} x)}{dx} = \frac{1}{1 + x^2} \qquad (4.50)$$

Integration

From the previous results, (4.48), (4.49) and (4.50), we have the following standard integrals

$$\int \frac{dx}{\sqrt{(1-x^2)}} = \sin^{-1} x + c \quad \text{or} \quad -\cos^{-1} x + c \tag{4.51}$$

$$\int \frac{dx}{1+x^2} = \tan^{-1} x + c \tag{4.52}$$

Example 4.5 *Find*

$$\int \frac{x^2}{1+x^2} \, dx$$

Put $x = \tan u$, then $dx/du = \sec^2 u$. Also $1 + x^2 = 1 + \tan^2 u = \sec^2 u$. Hence

$$\int \frac{x^2}{1+x^2} \, dx = \int \frac{(\tan^2 u)(\sec^2 u)}{\sec^2 u} \, du = \int \tan^2 u \, du$$

$$= \int (\sec^2 u - 1) \, du \quad \text{using identity (4.4)}$$

$$= \int \sec^2 u \, du - \int du$$

$$= \tan u - u = x - \tan^{-1} x$$

Hence

$$\int \frac{x^2}{1+x^2} \, dx = x - \tan^{-1} x$$

Hyberbolic functions

The so-called hyperbolic functions are very closely related to the trigonometric (or circular) functions. It is not intended to consider these functions in any detail in this book as they have little direct biological use, but, since they may arise in the evaluation of some integrals, a statement of their existence is necessary. The two basic forms are

$$\sinh x = \tfrac{1}{2}(e^x - e^{-x}) \tag{4.53}$$

pronounced 'shine x', and

$$\cosh x = \tfrac{1}{2}(e^x + e^{-x}) \qquad (4.54)$$

pronounced as it is written. We then have

$$\tanh x = \frac{\sinh x}{\cosh x} \qquad (4.55)$$

pronounced 'than x' or 'tansh x', and the reciprocal forms cosech x = 1/sinh x, sech x = 1/cosh x, and coth x = 1/tanh x, pronounced 'coshec', 'shec', and 'coth', respectively.

As an exercise, you may like to draw the graphs of these functions. Although, in the main, the graph of a hyperbolic function looks rather unlike its trigonometric counterpart (analogues are indicated by the name, e.g. sinh and sin) it is interesting to find that their mathematical properties are quite similar, but not identical. Thus, we have

$$\sin(x + y) = \sin x \cos y + \cos x \sin y$$

(4.6), and

$$\sinh(x + y) = \sinh x \cosh y + \cosh x \sinh y \qquad (4.56)$$

where $\sinh(x + y) = \tfrac{1}{2}(e^{x+y} - e^{-x-y})$; but whereas

$$\cos(x + y) = \cos x \cos y - \sin x \sin y$$

(4.7),

$$\cosh(x + y) = \cosh x \cosh y + \sinh x \sinh y \qquad (4.57)$$

Again, whereas $d(\sin x)/dx = \cos x$, and $d(\cos x)/dx = -\sin x$,

$$\frac{d(\sinh x)}{dx} = \cosh x \qquad (4.58)$$

as before, but

$$\frac{d(\cosh x)}{dx} = \sinh x \qquad (4.59)$$

Fourier series

In Chapter 5, pages 156–61, it is shown how several functions (of x) may be approximated, to any desired degree of accuracy, by polynomials of infinite degree of the form

$$f(x) = a_0 + a_1 x + a_2 x^2 + a_3 x^3 + \dots \qquad (4.60)$$

143

In (4.60) successive terms become smaller because as i increases the coefficients a_i decrease in size faster than values of x^i increase. Another way of writing (4.60) is

$$f(x) = \sum_{i=0}^{\infty} a_i x^i \qquad (4.61)$$

and when we approximate, using a finite number of terms,

$$f(x) \simeq \sum_{i=0}^{n} a_i x^i \qquad (4.62)$$

The approximation improves in accuracy as n, the number of terms included, increases.

For some purposes we require a series to approximate a function which recurs periodically; this can be achieved by a **Fourier series** which is a trigonometric function recurring every 2π units along the horizontal axis. A situation in which a quantity recurs in a cyclic manner, the pattern of occurrence of particular values of the quantity repeating themselves over and over again, may be analysed by a Fourier series; the period of the recurrence need not be restricted to the value 2π. Such problems are often found in data that consist of repeated measurements of the same item at regular intervals of time, say every year or every month or every day. These are called **time series**, and are often analysed by the technique of **spectral analysis**. A detailed treatment of these topics is beyond the scope of this book, but we can see the basic ideas underlying a Fourier series using a few simple examples (in which we shall use 2π as period).

Consider the series

$$\tfrac{1}{2} a_0 + a_1 \cos x + a_2 \cos 2x + \ldots + b_1 \sin x + b_2 \sin 2x + \ldots$$

$$= \tfrac{1}{2} a_0 + \sum_{r=1}^{\infty} (a_r \cos rx + b_r \sin rx) \qquad (4.63)$$

where r takes integer values and a_0, a_r, and b_r are constants. Since (4.63) is unchanged when x is replaced by $x + 2k$, where k is any integer, the series represents a periodic function of x with period 2π. Consequently it is sufficient to consider the behaviour of (4.63) in any interval of length 2π, e.g. $-\pi < x \leqslant \pi$ or $0 < x \leqslant 2\pi$.

Now let $f(x)$ be any function in, say, the interval $-\pi \leqslant x \leqslant \pi$; then if the coefficients in (4.63) are defined as

$$a_0 = \frac{1}{\pi} \int_{-\pi}^{\pi} f(x)\, \mathrm{d}x \qquad (4.64)$$

$$a_r = \frac{1}{\pi} \int_{-\pi}^{\pi} f(x)\cos rx\, \mathrm{d}x \qquad r = 1, 2, 3, \ldots \qquad (4.65)$$

$$b_r = \frac{1}{\pi} \int_{-\pi}^{\pi} f(x)\sin rx \, dx \qquad r = 1, 2, 3, \ldots \qquad (4.66)$$

the resulting series is called the Fourier series of $f(x)$, and the coefficients defined by (4.64), (4.65), and (4.66) are the **Fourier coefficients**. For most, but not all, functions we may write

$$f(x) = \tfrac{1}{2} a_0 + \sum_{r=1}^{\infty} (a_r \cos rx + b_r \sin rx) \qquad (4.67)$$

where the coefficients a_r and b_r are as defined in (4.64), (4.65) and (4.66), respectively.

Example 4.6   Find the Fourier series equal to $f(x) = x^2$ when

(a) $0 < x \leqslant 2\pi$, and (b) $-\pi < x \leqslant \pi$

(a)   Remembering that a Fourier series represents a *periodic* function, the full specification of $f(x)$ is

$$f(x) = (x - 2k\pi)^2 \qquad k = 0, 1, 2, \ldots$$
$$k = -1, \, -2, \, \ldots$$

such that $0 \leqslant (x - 2k\pi) \leqslant 2\pi$

and the graph of this function, in the range -2π to 6π, is shown in Figure 4.10a. Note the discontinuities every 2π units along the horizontal axis.

To evaluate the coefficients of (4.67), we first use (4.64):

$$a_0 = \frac{1}{\pi} \int_0^{2\pi} x^2 \, dx = \frac{1}{3\pi} [x^3]_0^{2\pi} = \frac{8\pi^2}{3} \qquad (4.68)$$

Now apply (4.65), giving

$$a_r = \frac{1}{\pi} \int_0^{2\pi} x^2 \cos rx \, dx$$

Integrating by parts, (3.2), and suppressing the constant of integration because in the end we are evaluating a definite integral:

$$\int x^2 \cos rx \, dx = x^2 \int \cos rx \, dx - \int \left[2x \int \cos rx \, dx \right] dx$$

$$= \frac{x^2}{r} \sin rx - \int \frac{2x}{r} \sin rx \, dx$$

(a)

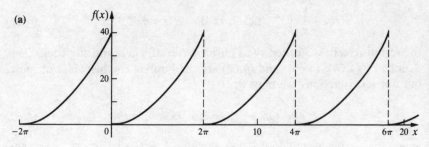

(b)

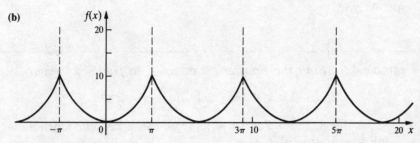

Figure 4.10 (a) Graph of the function $f(x) = (x - 2k\pi)^2$, $k = \ldots -2$, -1, 0, 1, $2, \ldots$, such that $0 \leqslant (x - 2k\pi) \leqslant 2\pi$. (b) Graph of the function $f(x) = (x - 2k\pi)^2$, $k = \ldots -2$, -1, 0, 1, $2, \ldots$, such that $-\pi \leqslant (x - 2k\pi) \leqslant \pi$.

So

$$\int x^2 \cos rx \, dx = \frac{x^2}{r} \sin rx - \frac{2}{r} \int x \sin rx \, dx \tag{4.69}$$

The remaining integral on the right-hand side of (4.69) also has to be integrated by parts:

$$\int x \sin rx \, dx = x \int \sin rx \, dx - \int \left[1 \int \sin rx \, dx \right] dx$$

$$= -\frac{x}{r} \cos rx - \int 1 \left(-\frac{1}{r} \cos rx \right) dx$$

$$= -\frac{x}{r} \cos rx + \frac{1}{r} \int \cos rx \, dx$$

$$= -\frac{x}{r} \cos rx + \frac{1}{r^2} \sin rx$$

So

$$\int x \sin rx \, dx = \frac{1}{r^2} \sin rx - \frac{x}{r} \cos rx \tag{4.70}$$

146

Substituting (4.70) back into (4.69) gives

$$\int x^2 \cos rx \, dx = \frac{x^2}{r} \sin rx - \frac{2}{r} \left(\frac{1}{r^2} \sin rx - \frac{x}{r} \cos rx \right)$$

i.e.

$$\int x^2 \cos rx \, dx = \frac{1}{r} \left(x^2 \sin rx + \frac{2x}{r} \cos rx - \frac{2}{r^2} \sin rx \right) \quad (4.71)$$

so

$$\int_0^{2\pi} x^2 \cos rx \, dx = \frac{1}{r} \left[x^2 \sin rx + \frac{2x}{r} \cos rx - \frac{2}{r^2} \sin rx \right]_0^{2\pi}$$

$$= \frac{4\pi}{r^2}$$

Hence

$$a_r = \frac{1}{\pi} \int_0^{2\pi} x^2 \cos rx \, dx = \frac{4}{r^2} \quad (4.72)$$

In an analagous way it can be shown that (Exercise 4.9 at end of this chapter)

$$b_r = \frac{1}{\pi} \int_0^{2\pi} x^2 \sin rx \, dx = -\frac{4\pi}{r} \quad (4.73)$$

Thus by substituting (4.68), (4.72), and (4.73) into (4.67), we have that

$$x^2 = \frac{4\pi^2}{3} + \sum_{r=1}^{\infty} \left(\frac{4}{r^2} \cos rx - \frac{4\pi}{r} \sin rx \right) \qquad 0 < x \leqslant 2\pi$$

or

$$x^2 = \frac{4\pi^2}{3} + 4 \sum_{r=1}^{\infty} \frac{1}{r^2} \cos rx - 4\pi \sum_{r=1}^{\infty} \frac{1}{r} \sin rx \qquad 0 < x \leqslant 2\pi \quad (4.74)$$

In practice, the accuracy of the series is determined by the number of pairs of terms (each pair involving a cosine term and a sine term) included in the summations, and Figure 4.11 shows the results of using 1, 2, 5, and 10 pairs of terms. In all cases there is a marked deviation away from the curve of $y = x^2$ near the ends of the 2π interval; this is because at $x = 0, 2\pi, \ldots$ the value of the Fourier series is half way between 0 and $(2\pi)^2$ ($\simeq 39.4784$).

(b) The graph of the function in the range $-\pi$ to π appears in Figure 4.10b. The first constant, a_0, is evaluated directly as

$$a_0 = \frac{1}{\pi} \int_{-\pi}^{\pi} x^2 \, dx = \frac{1}{3\pi} \left[x^3 \right]_{-\pi}^{\pi} = \frac{1}{3\pi} (2\pi^3) = \tfrac{2}{3} \pi^2 \quad (4.75)$$

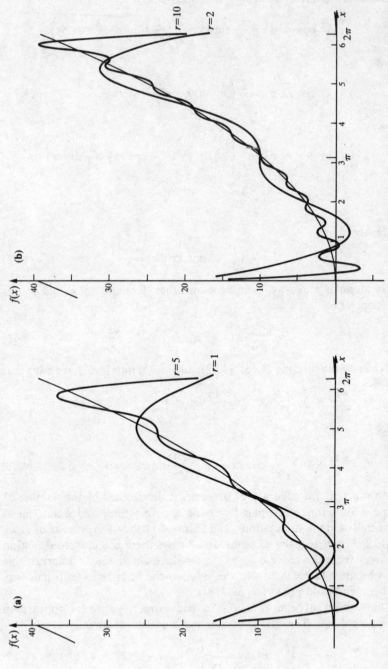

Figure 4.11 Graph of the function $f(x) = x^2$ in the range $0 \leqslant x \leqslant 2\pi$, together with Fourier series equivalent curves of equation (4.74) with $r = 1$ and 5 (a), and $r = 2$ and 10 (b).

To obtain the value of a_r $(r \neq 0)$, we use the result in (4.71) and employ the limits $-\pi$ and π. Thus

$$\int_{-\pi}^{\pi} x^2 \cos rx \, dx = \frac{1}{r}\left[x^2 \sin rx + \frac{2x}{r}\cos rx - \frac{2}{r^2}\sin rx\right]_{-\pi}^{\pi}$$

$$= \frac{1}{r}\left\{\frac{2\pi}{r}(-1)^r - \frac{2(-\pi)}{r}(-1)^r\right\}$$

$$= \frac{4\pi}{r^2}(-1)^r$$

(See page 159 for an explanation of the term $(-1)^r$.) Hence

$$a_r = \frac{1}{\pi}\int_{-\pi}^{\pi} x^2 \cos rx \, dx = \frac{4}{r^2}(-1)^r \qquad (4.76)$$

Similarly, it can be shown that

$$\int_{-\pi}^{\pi} x^2 \sin rx \, dx = \frac{1}{r}\left[-x^2\cos rx + \frac{2x}{r}\sin rx + \frac{2}{r^2}\cos rx\right]_{-\pi}^{\pi}$$

$$= \frac{1}{r}\left[\left\{-\pi^2(-1)^r + \frac{2}{r^2}(-1)^r\right\}\right.$$

$$\left. - \left\{-(-\pi^2)(-1)^r + \frac{2}{r^2}(-1)^r\right\}\right]$$

$$= 0$$

Hence

$$b_r = 0. \qquad (4.77)$$

Thus by substituting (4.75), (4.76), and (4.77) into (4.67), we have that

$$x^2 = \frac{\pi^2}{3} + 4\sum_{r=1}^{\infty}\frac{(-1)^r}{r^2}\cos rx \qquad -\pi < x \leqslant \pi \qquad (4.78)$$

The result of (4.74) yields the sums of two infinite series which cannot easily be obtained in any other way. Put $x = 0$, then

$$2\pi^2 = \frac{4\pi^2}{3} + 4\sum_{r=1}^{\infty}\frac{1}{r^2}$$

since $\cos 0 = 1$, $\sin 0 = 0$, and $\frac{1}{2}(2\pi)^2$ is the average value at the discontinuity when $x = 0$. Rearranging, we obtain

$$\sum_{r=1}^{\infty}\frac{1}{r^2} = \frac{\pi^2}{6} \qquad (4.79)$$

Next, put $x = \pi$, then (4.74) becomes

$$\pi^2 = \frac{4\pi^2}{3} + 4 \sum_{r=1}^{\infty} \frac{1}{r^2} (-1)^r$$

since $\sin r\pi = 0$ for all integer values of r, and $\cos r\pi = 1$ if r is even and -1 if r is odd. So, finally,

$$\sum_{r=1}^{\infty} \frac{(-1)^r}{r^2} = -\frac{\pi}{12} \tag{4.80}$$

Hopefully, this short introduction to Fourier series will have given you some feel for the subject and should enable you to read the more standard mathematical texts such as Heading (1970). The application of Fourier series to the spectral analysis of time series data is described in Chatfield (1984).

Suggestions for further reading

The topics in this chapter are covered in the general mathematical texts referenced elsewhere in this book.

The application of Fourier series in spectral analysis is described by **Chatfield, C.** (1984). *The analysis of time series: an introduction*, 3rd edn. London: Chapman & Hall.

One field of application of trigonometric functions, and double integrals, is in the modelling of light interception by single plants and foliage canopies. Two relevant papers are **Charles-Edwards, D. A.** and **J. H. M. Thornley** (1973). Light interception by an isolated plant. A simple model. *Annals of Botany* **37**, 919–28, and **Norman, J. M.** and **J. M. Welles** (1983). Radiative transfer in an array of canopies. *Agronomy Journal* **75**, 481–8.

Exercises

4.1 Convert the following angles in degrees to radians:
   (a) $60°$ (b) $100°$ (c) $15°$

4.2 Convert the following angles in radians to degrees:
   (a) 0.1 rad   (b) 2.7 rad   (c) 1.4 rad

4.3 Differentiate with respect to x:
   (a) $x^2 \sin 2x$ (b) $x/(\tan x)$ (c) $\cos(2 - 3x)$

4.4 Find the co-ordinates of the maximum and minimum points on the curve of the function $y = 50e^{0.2 \sin x}$ in the range $0 \leqslant x \leqslant 2\pi$.

EXERCISES

4.5 Find the following integrals, using the suggested substitutions:
   (a) $\int(\sin 4x)\,dx: z = 4x$ (b) $\int x e^{-x^2}\,dx: z = x^2$
   (c) $\int(\sin x \cos x)\,dx: z = \sin x$

4.6 Integrate by parts:
   (a) $\int x \cos x\,dx$ (b) $\int e^x \sin x\,dx$

4.7 Differentiate with respect to x:
   (a) $\sin^{-1}(2x-1)$ (b) $\cos^{-1}(1-x)$ (c) $\tan^{-1} 3x$

4.8 Find

   (a) $\int \dfrac{dx}{\sqrt{(4-x^2)}}$ (b) $\int \dfrac{dx}{\sqrt{(9-4x^2)}}$

4.9 Show that

$$b_r = \frac{1}{\pi} \int_0^{2\pi} x^2 \sin rx\,dx = -\frac{4\pi}{r}$$

4.10 Find the Fourier series equal to $f(x) = e^x$ in the interval $-\pi < x \leqslant \pi$.

5

A miscellany: series, complex numbers, partial fractions and numerical analysis

The diverse topics forming this chapter have no *direct* biological uses but provide certain essential background material for many other subjects in the book. Thus, for example, complex numbers arise in the solution of some differential equations; partial fractions are a central feature in the use of the Laplace transform to solve linear differential equations; mathematical series form a basis in considering small errors and in the evaluation of many kinds of mathematical function; and some of the principal ideas of numerical analysis should be appreciated by any user of mathematical methods.

The first part of this chapter is essentially a continuation of Chapter 8 in Causton (1983); indeed the first section here continues exactly where that chapter finished – an extension of the binomial series to the cases where the index is negative and/or fractional. We shall next introduce two related and very powerful series, namely, the Maclaurin and Taylor series, and then show some of their many uses.

The binomial series with negative and fractional indices

The expansion of $(a + x)^n$ is given by

$$(a + x)^n = a^n + na^{n-1}x + \frac{n(n-1)}{2!} a^{n-2}x^2 + \frac{n(n-1)(n-2)}{3!} a^{n-3}x^3$$

$$+ \ldots + \frac{n(n-1)(n-2)\ldots(n-i+2)}{(i-1)!} a^{n-i+1}x^{i-1} + \ldots \quad (5.1)$$

If n is a positive integer the resulting number of terms is $n + 1$; for any other value of n the number of terms is infinite.

Example 5.1 *Find the first four terms in the expansions of*

$$(a)\ \frac{1}{(2+x)^3} \qquad (b)\ \sqrt{(4+5x)} \qquad (c)\ \frac{1}{\sqrt[3]{(8-3x)}}$$

(a)

$$\frac{1}{(2+x)^3} = (2+x)^{-3}$$

In (5.1), put $a = 2$ and $n = -3$; then

$$(2+x)^{-3} = (2)^{-3} + (-3)(2)^{-4}x + \frac{(-3)(-4)}{(1)(2)}(2)^{-5}x^2$$
$$+ \frac{(-3)(-4)(-5)}{(1)(2)(3)}(2)^{-6}x^3 + \ldots$$

$$= \frac{1}{8} + \frac{(-3)x}{16} + \frac{12x^2}{64} + \frac{(-60)x^3}{144} + \ldots$$

i.e.

$$(2+x)^{-3} = \tfrac{1}{8} - \tfrac{3}{16}x + \tfrac{3}{16}x^2 - \tfrac{5}{12}x^3 + \ldots$$

(b)

$$\sqrt{(4+5x)} = (4+5x)^{\frac{1}{2}}$$

In (5.1), put $a = 4$, $x = 5x$, and $n = \frac{1}{2}$; then

$$(4+5x)^{\frac{1}{2}} = (4)^{\frac{1}{2}} + (1/2)(4)^{-\frac{1}{2}}(5x) + \frac{(1/2)(-1/2)}{(1)(2)}(4)^{-\frac{3}{2}}(5x)^2$$

$$+ \frac{(1/2)(-1/2)(-3/2)}{(1)(2)(3)}(4)^{-\frac{5}{2}}(5x)^3 + \ldots$$

$$= \sqrt{4} + \frac{5}{2\sqrt{4}}x + \frac{(25)(-1/4)}{2\sqrt{(4)^3}}x^2 + \frac{(125)(3/8)}{6\sqrt{(4)^5}}x^3 + \ldots$$

$$= 2 + \frac{5}{2\sqrt{4}}x - \frac{25}{8\sqrt{64}}x^2 + \frac{375}{48\sqrt{1024}}x^3 + \ldots$$

Hence

$$\sqrt{(4+5x)} = 2 + \tfrac{5}{4}x - \tfrac{25}{64}x^2 + \tfrac{125}{512}x^3 + \ldots$$

(c)

$$\frac{1}{\sqrt[3]{(8-3x)}} = \frac{1}{(8-3x)^{\frac{1}{3}}} = (8-3x)^{-\frac{1}{3}}$$

153

In (5.1), put $a = 8$, $x = -3x$, and $n = -\frac{1}{3}$; then

$$\{8 + (-3x)\}^{-\frac{1}{3}} = (8)^{-\frac{1}{3}} + (-1/3)(8)^{-\frac{4}{3}}(-3x)$$

$$+ \frac{(-1/3)(-4/3)}{(1)(2)} (8)^{-\frac{7}{3}}(-3x)^2$$

$$+ \frac{(-1/3)(-4/3)(-7/3)}{(1)(2)(3)} (8)^{-\frac{10}{3}}(-3x)^3 + \ldots$$

$$= (8)^{-\frac{1}{3}} + \frac{(-1/3)(-3x)}{(8)^{\frac{4}{3}}} + \frac{(-1/3)(-4/3)(-3x)^2}{(1)(2)(8)^{\frac{7}{3}}}$$

$$+ \frac{(-1/3)(-4/3)(-7/3)(-3x)^3}{(1)(2)(3)(8)^{\frac{10}{3}}} + \ldots$$

$$= \frac{1}{(8)^{\frac{1}{3}}} + \frac{1}{(8)^{\frac{4}{3}}} x + \frac{4}{(2)(8)^{\frac{7}{3}}} x^2 + \frac{28}{(6)(8)^{\frac{10}{3}}} x^3 + \ldots$$

$$= \frac{1}{\sqrt[3]{8}} + \frac{1}{\sqrt[3]{(8)^4}} x + \frac{2}{\sqrt[3]{(8)^7}} x^2 + \frac{14}{3\sqrt[3]{(8)^{10}}} x^3 + \ldots$$

$$= \frac{1}{2} + \frac{1}{16} x + \frac{2}{128} x^2 + \frac{14}{(3)(1024)} x^3 + \ldots$$

Thus

$$\frac{1}{\sqrt[3]{(8 - 3x)}} = \frac{1}{2} + \frac{1}{16} x + \frac{1}{64} x^2 + \frac{7}{1536} x^3 + \ldots$$

The Maclaurin and Taylor series

In the expansion of $(a + x)^n$, whatever the value of n, we found a series of terms which can be regarded as a polynomial in x, since successive terms contain successive integer powers of x. The binomial series polynomial has a finite number of terms only if n is a positive integer, otherwise there are an infinite number of terms. Many important mathematical functions can be defined in terms of what might be called a polynomial of infinite degree. For instance,

$$e^x = 1 + x + \frac{x^2}{2!} + \frac{x^3}{3!} + \ldots$$

thus enabling e^x to be calculated for any value of x.

The two series described in this section are closely related to one another:

indeed, the Maclaurin series is merely a simpler special case of the Taylor series, but we shall describe the former separately first.

The Maclaurin series

Suppose we have a function of x, $f(x)$, that we assume can be defined in terms of a polynomial of infinite degree, i.e.

$$f(x) \equiv a_0 + a_1 x + a_2 x^2 + a_3 x^3 + a_4 x^4 + \dots \qquad (5.2)$$

We have used the sign $\equiv$ instead of $=$ to show that relationship (5.2) holds for *all* values of x. Now differentiate (5.2) successively with respect to x:

$$f'(x) \equiv a_1 + 2a_2 x + 3a_3 x^2 + 4a_4 x^3 + \dots \qquad (5.3)$$

$$f''(x) \equiv 2a_2 + 6a_3 x + 12a_4 x^2 + \dots \qquad (5.4)$$

$$f'''(x) \equiv 6a_3 + 24a_4 x + \dots \qquad (5.5)$$

$$f^{IV}(x) \equiv 24a_4 + \dots \qquad (5.6)$$

Put $x = 0$; then

$$f(0) = a_0 \quad \text{(from 5.2)} \qquad f'(0) = a_1 \quad \text{(from 5.3)}$$

$$f''(0) = 2a_2 \quad \text{(from 5.4)} \qquad f'''(0) = 6a_3 \quad \text{(from 5.5)}$$

$$f^{IV}(0) = 24a_4 \quad \text{(from 5.6)}$$

and so on. Therefore

$$a_0 = f(0), \quad a_1 = f'(0), \quad a_2 = f''(0)/2!, \quad a_3 = f'''(0)/3!, \quad a_4 = f^{IV}(0)/4!,$$

and generally

$$a_r = f^{(r)}(0)/r!$$

for the rth term, where $f^{(r)}(0)$ stands for the rth derivative of the function evaluated at $x = 0$.

At this stage we have evaluated the coefficients on the right-hand side of (5.2) when $x = 0$; but since (5.2) holds for all values of x, the values of the coefficients are the same regardless of the magnitude of x. So, substituting back into (5.2), we have

$$f(x) \equiv f(0) + f'(0)x + \frac{f''(0)}{2!} x^2 + \frac{f'''(0)}{3!} x^3 + \dots \qquad (5.7)$$

or

$$f(x) = \sum_{r=0}^{\infty} \frac{f^{(r)}(0)}{r!} x^r \qquad (5.8)$$

Expressions (5.7) and (5.8) are different ways of writing down the Maclaurin expansion of $f(x)$.

The Taylor series

We derive this series in a manner identical to that for the Maclaurin series. Suppose we have a function of h, of the form $f(x + h)$, that we assume can be defined in terms of

$$f(x + h) \equiv a_0 + a_1 h + a_2 h^2 + a_3 h^3 + a_4 h^4 = \ldots \qquad (5.9)$$

Differentiate (5.9) successively with respect to h:

$$f'(x + h) \equiv a_1 + 2a_2 h + 3a_3 h^2 = 4a_4 h^3 + \ldots \qquad (5.10)$$

$$f''(x + h) \equiv \quad\quad 2a_2 + 6a_3 h + 12a_4 h^2 + \ldots \qquad (5.11)$$

$$f'''(x + h) \equiv \quad\quad\quad 6a_3 + 24a_4 h + \ldots \qquad (5.12)$$

$$f^{IV}(x + h) \equiv \quad\quad\quad\quad 24a_4 + \ldots \qquad (5.13)$$

Hence, as before, with $h = 0$:

$$a_0 = f(x), \qquad a_1 = f'(x), \qquad a_2 = f''(x)/2!,$$

$$a_3 = f'''(x)/3!, \qquad a_4 = f^{IV}(x)/4!$$

and, generally,

$$a_r = f^{(r)}(x)/r!$$

for the rth term, where $f^{(r)}(x)$ stands for the rth derivative of the function evaluated at $h = 0$.

As before, with the Maclaurin series, we have evaluated the coefficients on the right-hand side of (5.9) when $h = 0$; but since (5.9) holds for all values of h, the coefficients are the same regardless of the value of h. So, substituting back into (5.9), we have

$$f(x + h) \equiv f(x) + f'(x)h + \frac{f''(x)}{2!} h^2 + \frac{f'''(x)}{3!} h^3 + \ldots \qquad (5.14)$$

or

$$f(x + h) = \sum_{r=0}^{\infty} \frac{f^{(r)}(x)}{r!} h^r \qquad (5.15)$$

Expressions (5.14) and (5.15) are the two different ways of writing down the Taylor expansion of $f(x + h)$.

Applications of the Maclaurin and Taylor series

The primary application of Maclaurin and Taylor series is in the expression of various kinds of function in terms of a polynomial of infinite degree.

There are other important applications also: we shall examine a range of applications in this section.

The exponential function

The exponential function is defined as that function, $f(x)$, which has the property

$$f'(x) = f(x)$$

Hence $f''(x) = f(x)$, $f'''(x) = fx$, ...; or, put another way, $f(x)$ and all its derivatives are given by e^x. Now $f(0) = e^0 = 1$. Hence, substituting into (5.7), we have

$$e^x = 1 + x + \frac{x^2}{2!} + \frac{x^3}{3!} + \ldots$$

which is the required result.

Trigonometric functions

If an angle x is measured in radians, it is possible to obtain the trigonometric functions $\sin x$ and $\cos x$ as the sums of infinite series. This enables one to find any trigonometric function of any angle if tables or a scientific calculator are not available, although the arithmetic work would be tedious. In fact, scientific calculators evaluate trigonometric functions by using a sufficient finite number of terms of these series expansions to obtain the accuracy required.

THE SINE FUNCTION

Putting $f(x) = \sin x$, we have $f'(x) = \cos x$, $f''(x) = -\sin x$, $f'''(x) = -\cos x$, $f^{IV}(x) = \sin x$, ... (eqns (4.27) and (4.30)). Furthermore, $\sin 0 = 0$ and $\cos 0 = 1$; so, substituting into (5.7) gives

$$\sin x = \sin 0 + (\cos 0)x + \frac{(-\sin 0)}{2!} x^2 + \frac{(-\cos 0)}{3!} x^3$$

$$+ \frac{\sin 0}{4!} x^4 + \frac{\cos 0}{5!} x^5 + \frac{(-\sin 0)}{6!} x^6 + \frac{(-\cos 0)}{7!} x^7 \ldots$$

i.e.

$$\sin x = x - \frac{x^3}{3!} + \frac{x^5}{5!} - \frac{x^7}{7!} + \ldots \tag{5.16}$$

THE COSINE FUNCTION

By a similar argument, which is left to the reader (see Exercise 5.2 at the end of the chapter), we have

$$\cos x = 1 - \frac{x^2}{2!} + \frac{x^4}{4!} - \frac{x^6}{6!} + \ldots \tag{5.17}$$

In both of the above functions, as with the exponential series, the smaller x is the more rapidly does the series converge.

Example 5.2   Find, from the appropriate series expansion:

    (a) sin 35°   *(b)* cos 21°12' to four decimal places

(a) First, the angle must be expressed in radians. From (4.2) it is evident that 35° is equal to $35\pi/180 \simeq 0.61$ rad. Hence

$$\sin 0.61 = 0.61 - \frac{(0.61)^3}{3!} + \frac{(0.61)^5}{5!} - \frac{(0.61)^7}{7!} + \ldots$$

$$= 0.61 - 0.037\ 830 + 0.000\ 704 - 0.000\ 006$$

$$= 0.572\ 868$$

Thus sin 35° $\simeq 0.5729$ correct to four decimal places. The $\simeq$ sign is still used here despite the specification of accuracy to four decimal places because of a small error of conversion from degrees to radians to simplify the arithmetic. In fact, 35° $= 0.610\ 865\ 24$ rad correct to eight decimal places and sin 35° $= 0.5736$ correct to four decimal places.

(b) The angle 21°12' is equal to to 21.2° and in radians is 0.370 009 80 to eight decimal places. The approximate conversion of 21°12' to 0.37 rad is therefore a much closer one than in (a) above. So

$$\cos 0.37 = 1 - \frac{(0.37)^2}{2!} + \frac{(0.37)^4}{4!} - \frac{(0.37)^6}{6!} + \ldots$$

$$= 1 - 0.068\ 450 + 0.000\ 781 - 0.000\ 004$$

$$= 0.932\ 327$$

Thus cos 21°12' $= 0.9323$ correct to four decimal places which is, in fact, exact to four decimal places.

The logarithmic function

It is not possible to expand $\log_e x$ as an infinite series directly, but $\log_e(1 + x)$ can be so expressed when $|x| < 1$ using the Taylor expansion.

In order to match the notation with that of (5.14), let us specify the function as $\log_e(1 + h)$, then x in (5.14) is equal to 1 in the current function. We require the successive derivatives of $\log_e(1 + h)$, and these are

$$\frac{d\{\log_e(1 + h)\}}{dh} = \frac{1}{1 + h}$$

$$\frac{d^2\{\log_e(1 + h)\}}{dh^2} = -\frac{1}{(1 + h)^2}$$

$$\frac{d^3\{\log_e(1 + h)\}}{dh^3} = \frac{2(1 + h)}{(1 + h)^4} = \frac{2!}{(1 + h)^3}$$

$$\frac{d^4\{\log_e(1 + h)\}}{dh^4} = -\frac{2!3(1 + h)^2}{(1 + h)^6} = -\frac{3!}{(1 + h)^4}$$

$$\frac{d^5\{\log_e(1 + h)\}}{dh^5} = \frac{3!4(1 + h)^3}{(1 + h)^8} = \frac{4!}{(1 + h)^5}$$

and, generally,

$$\frac{d^r\{\log_e(1 + h)\}}{dh^r} = (-1)^{r-1}\frac{(r - 1)!}{(1 + h)^r}$$

The prefix term $(-1)^{r-1}$ ensures that when r is odd the derivative is positive and when r is even the derivative is negative, without altering the absolute value of the derivative.

When $h = 0$, the expression $\log_e(1 + h)$ and its consecutive derivatives are: 0 (because $\log_e 1 = 0$), 1, -1, 2!, $-3!$, 4! ... Substituting into (5.14) gives

$$\log_e(1 + h) = 0 + h - \frac{h^2}{2!} + \frac{2!h^3}{3!} - \frac{3!h^4}{4!} + \cdots$$

i.e.

$$\log_e(1 + h) = h - \frac{h^2}{2} + \frac{h^3}{3} - \frac{h^4}{4} \cdots \qquad (5.18)$$

Now put $h = -h$ in (5.18), then

$$\log_e\{1 + (-h)\} = (-h) - \frac{(-h)^2}{2} + \frac{(-h)^3}{3} - \frac{(-h)^4}{4} + \cdots$$

i.e.

$$\log_e(1 - h) = -h - \frac{h^2}{2} - \frac{h^3}{3} - \frac{h^4}{4} - \cdots \qquad (5.19)$$

The relationships (5.18) and (5.19) are valid only when $|h| < 1$ (i.e.

159

$-1 < h < 1$), because it is only when this condition holds that the right-hand sides converge. The nearer h is to zero, the faster the convergence. Hence the utility of these series in evaluating natural logarithms is very limited. Can we obtain from (5.18) and (5.19) a series which will enable the natural logarithm of *any* number to be evaluated? Put

$$x = \frac{1+h}{1-h} \qquad (5.20)$$

then by cross-multiplication and rearrangement, we obtain

$$h = \frac{x-1}{x+1} \qquad (5.21)$$

From (5.21) it can be seen that whatever the value of x may be, h will always be in the range $-1 < h < 1$. If, then, we can find a series for $\log_e x$ derived from (5.18) and (5.19), it will be valid for any value of x. Now from (5.20)

$$\log_e x = \log_e\left(\frac{1+h}{1-h}\right) = \log_e(1+h) - \log_e(1-h)$$

$$= \left(h - \frac{h^2}{2} + \frac{h^3}{3} - \frac{h^4}{4} + \ldots\right) - \left(-h - \frac{h^2}{2} - \frac{h^3}{3} - \frac{h^4}{4} - \ldots\right)$$

$$= h - \frac{h^2}{2} + \frac{h^3}{3} - \frac{h^4}{4} + \ldots + h + \frac{h^2}{2} + \frac{h^3}{3} + \frac{h^4}{4} + \ldots$$

$$= 2h + 2\frac{h^3}{3} + 2\frac{h^5}{5} + \ldots \text{(since all terms in even}$$
$$\text{powers of } h \text{ cancel out)}$$

Hence

$$\log_e x = 2\left(h + \frac{h^3}{3} + \frac{h^5}{5} + \ldots\right)$$

But from (5.21), $h = (x-1)/(x+1)$, so

$$\log_e x = 2\left\{\frac{x-1}{x+1} + \frac{1}{3}\left(\frac{x-1}{x+1}\right)^3 + \frac{1}{5}\left(\frac{x-1}{x+1}\right)^5 + \ldots\right\} \qquad (5.22)$$

Example 5.3 Evaluate $\log_e 3$ correct to four decimal places.

Put $x = 3$ in (5.22); then

$$\log_e 3 = 2\left\{\frac{3-1}{3+1} + \frac{1}{3}\left(\frac{3-1}{3+1}\right)^3 + \frac{1}{5}\left(\frac{3-1}{3+1}\right)^7 + \ldots\right\}$$

$$= 2\left\{\frac{2}{4} + \frac{1}{3}\left(\frac{2}{4}\right)^3 + \frac{1}{5}\left(\frac{2}{4}\right)^5 + \frac{1}{7}\left(\frac{2}{4}\right)^7 + \frac{1}{9}\left(\frac{2}{4}\right)^9 + \frac{1}{11}\left(\frac{2}{4}\right)^{11} + \ldots\right\}$$

$$= 2\left\{\frac{1}{2} + \left(\frac{1}{3}\right)\left(\frac{1}{8}\right) + \left(\frac{1}{5}\right)\left(\frac{1}{32}\right) + \left(\frac{1}{7}\right)\left(\frac{1}{128}\right) + \left(\frac{1}{9}\right)\left(\frac{1}{512}\right)\right.$$

$$\left. + \left(\frac{1}{11}\right)\left(\frac{1}{2048}\right) + \ldots\right\}$$

$$= 2\left(\frac{1}{2} + \frac{1}{24} + \frac{1}{160} + \frac{1}{896} + \frac{1}{4608} + \frac{1}{22\,528} + \ldots\right)$$

$$= 2(0.5 + 0.041\,67 + 0.006\,250 + 0.001\,116$$
$$+ 0.000\,217\,0 + 0.000\,044\,39 + \ldots)$$

$$= 2 \times 0.549\,297\,3 = 1.098\,594\,78$$

So

$$\log_e 3 = 1.0986 \text{ correct to four decimal places}$$

Small increments

Refer to the situation in Figure 5.1. Point P is any point on the curve $y = f(x)$, and point Q is also a point on the curve very close to P so tht δx is small. Now if x, δx, and $f(x)$ are known, the y-co-ordinate of point Q, $f(x + \delta x)$, can be calculated exactly; but for some applications it is more useful to express the result for $f(x + \delta x)$ in a different way.

Now from Taylor's expansion, (5.14), we have

$$f(x + \delta x) = f(x) + f'(x)\,\delta x + \frac{f''(x)}{2!}\,(\delta x)^2 + \frac{f'''(x)}{3!}\,(\delta x)^3 + \ldots \qquad (5.23)$$

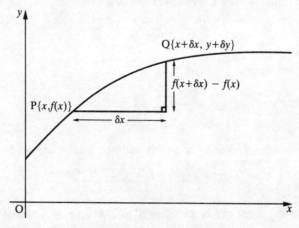

Figure 5.1 The geometrical basis of small increments.

161

Since δx is small, another way of writing (5.23) is

$$f(x + \delta x) = f(x) + f'(x)\,\delta x + O\{(\delta x)^2\} \tag{5.24}$$

where $O\{(\delta x)^2$ is read as 'order of delta x squared' and embraces the $(\delta x)^2$ term and all terms beyond it to the right in expression (5.23). Again, since δx is small, we have the approximate relationship:

$$f(x + \delta x) \simeq f(x) + f'(x)\,\delta x \tag{5.25}$$

and the smaller δx is, the more accurate expression (5.25) becomes.

Further, since $y = f(x)$ and $y + \delta y = f(x + \delta x)$, then the vertical increment in Figure 5.1 is in fact δy; so

$$\delta y = f(x + \delta x) - f(x)$$

and so from (5.25) we have

$$\delta y \simeq f'(x)\,\delta x \tag{5.26}$$

or, more accurately,

$$\delta y = f'(x)\,\delta x + O\{(\delta x)^2\} \tag{5.27}$$

In many books, relationship (5.26) is written as

$$dy = f'(x)\,dx \tag{5.28}$$

where dx and dy are known as **differentials**. Technically dx and dy used in this way are quite distinct from the same symbols in the term dy/dx in that the former pair are not vanishingly small. In fact, expression (5.28) is highly confusing in more ways than one and its use should be avoided. If you come across (5.28) in other publications, translate it either in your mind or on paper into either of the expressions (5.26) or (5.27).

Complex numbers

Consider the quadratic equation

$$z^2 + 2z + 5 = 0$$

Using the usual formula to solve a quadratic equation, we have

$$z = \frac{-2 \pm \sqrt{(4 - 20)}}{2} = \frac{-2 \pm \sqrt{(-16)}}{2}$$

The square root term can be partitioned as $\sqrt{\{(16)(-1)\}}$, giving the solution

either $z = -1 + 4\sqrt{(-1)}$ or $z = -1 - 4\sqrt{(-1)}$

Since $\sqrt{(-1)}$ is not a real number, the compound numbers $z = -1 \pm 4\sqrt{(-1)}$ are not real either. The quantity $\sqrt{(-1)}$ is quite literally an imaginary number, and it is always given the symbol i; the number is defined as

$$i^2 = -1$$

Numbers such as z above, consisting of both real numbers and i, are called **complex numbers**; they take the form

$$z = x \pm iy$$

where x and y are real numbers. For a complex number $z = x + iy$, $z^* = x - iy$ is called the **complex conjugate** of z. Further, the real number x, standing on its own, is known as the real part of z (Rl z); and the real number y, which multiplies i, is called the imaginary part of z (Im z). Thus if $z = 2 - 3i$, Rl $z = 2$ and Im $z = -3$.

The Argand diagram

Just as a real number may be represented on a horizontal scale or straight line (the real line), so may a complex number be represented on a two-dimensional graph (the complex z-plane). In Figure 5.2 the complex number $z = x + iy$ is represented by point P whose Cartesian co-ordinates are (x, y). The x- and y-axes may be called the real and imaginary axes, respectively. If y is 0, z is real and lies on the x-axis; similarly, if x is zero, z is a **pure imaginary** and lies on the y-axis. The Cartesian co-ordinates of i are $(0, 1)$.

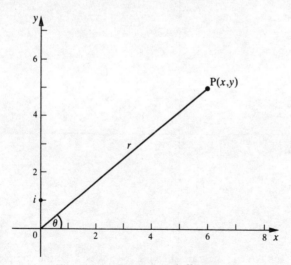

Figure 5.2 The Argand diagram.

163

The **modulus** of z is the length of OP, r (always positive), and we write

$$r = |z| = \sqrt{(x^2 + y^2)}$$

The angle θ, measured in radians, between the x-axis and OP in anti-clockwise direction is called the **argument** of z and is written as

$$\theta = \arg z = \tan^{-1}(y/x)$$

Example 5.4   Write down the modulus and argument of each of the following complex numbers. Also plot the numbers on an Argand diagram.

(a) $3i$ *(b)* $4 + 4i$ *(c)* $4 + 3i$ *(d)* $4 - 2i$ *(e)* $1 + i\sqrt{3}$

(a)   This is a pure imaginary number lying on the y-axis, so the modulus is $\sqrt{3^2} = 3$ and the argument is $\tfrac{1}{2}\pi$.

(b) $|z| = \sqrt{32} \simeq 5.6569$ $\theta = \tan^{-1}(4/4) = \tfrac{1}{4}\pi$

(c) $|z| = \sqrt{(16 + 9)} = 5$ $\theta = \tan^{-1}(3/4) \simeq 0.6435$

(d) $|z| = \sqrt{(16 + 4)} \simeq 4.4721$ $\theta = \tan^{-1}(-2/4)$
which implies that the point P is in the lower right-hand quadrant. The angle θ can be said to lie either between $\tfrac{3}{2}\pi$ and 2π or between 0 and

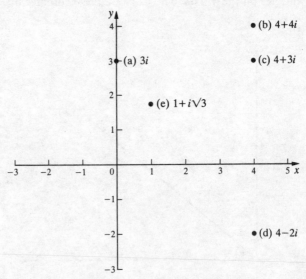

Figure 5.3   The Argand diagram showing the positions of the complex numbers in Example 5.4.

$-\frac{1}{2}\pi$. It is conventional to choose as the principal value of the argument the angle lying in the range $-\pi < \arg z \leqslant \pi$. Hence our present argument is negative and is -0.4637 approximately.

(e) $|z| = \sqrt{(1+3)} = 2$ $\theta = \tan^{-1}\sqrt{3} = \frac{1}{3}\pi$

The Argand diagram, showing each of these complex numbers, is exhibited in Figure 5.3.

Complex arithmetic

We have already defined i in the form $i^2 = -1$, and we now also assume that i obeys the rules of ordinary algebra. You should check the following results:

$$i^2 = -1 \qquad i^3 = -i \qquad i^4 = 1 \qquad i^5 = i$$
$$i^6 = -1 \qquad i^7 = -i \qquad i^8 = 1 \qquad i^9 = i \text{ etc.}$$

EQUALITY
For two complex numbers, $z_1 = x_1 + iy_1$ and $z_2 = x_2 + iy_2$, to be equal to one another means that $x_1 = x_2$ and $y_1 = y_2$. Thus for

$$z_1 = z_2 = z \text{ (say)}$$

implies that $x_1 = x_2 = x$ (say) and $y_1 = y_2 = y$ (say). A complex equation is really *two* equations – one for the real parts and one for the imaginary parts of the complex numbers involved.

ADDITION
The arithmetic of complex numbers follows the same laws as for real numbers. For addition, we have

$$z_1 + z_2 = x_1 + iy_1 + x_2 + iy_2$$
$$= (x_1 + x_2) + i(y_1 + y_2)$$

which is still a complex number.

SUBTRACTION
The procedure for subtraction of complex numbers is identical to that for addition:

$$z_1 - z_2 = x_1 + iy_1 - x_2 - iy_2$$
$$= (x_1 - x_2) + i(y_1 - y_2)$$

We may unify both addition and subtraction of complex numbers in one

165

equation, i.e.

$$z_1 \pm z_2 = (x_1 \pm x_2) + i(y_1 \pm y_2) \tag{5.29}$$

MULTIPLICATION

Again, by the usual laws of algebra we have

$$z_1 z_2 = (x_1 + iy_1)(x_2 + iy_2)$$

$$= x_1 x_2 + ix_1 y_2 + ix_2 y_1 + i^2 y_1 y_2$$

But $i^2 = -1$, so with rearrangement:

$$z_1 z_2 = (x_1 x_2 - y_1 y_2) + i(x_1 y_2 + x_2 y_1) \tag{5.30}$$

still a complex number.

DIVISION

We require an expression for the quotient

$$\frac{z_1}{z_2} = \frac{x_1 + iy_1}{x_2 + iy_2}$$

Multiply both numerator and denominator by the complex conjugate (see page 163) of the denominator:

$$\frac{z_1}{z_2} = \frac{(x_1 + iy_1)(x_2 - iy_2)}{(x_2 + iy_2)(x_2 - iy_2)}$$

$$= \frac{x_1 x_2 - ix_1 y_2 + ix_2 y_1 - i^2 y_1 y_2}{x_2^2 - ix_2 y_2 + ix_2 y_2 - i^2 y_2^2}$$

Again, remembering that $i^2 = -1$, we have

$$\frac{z_1}{z_2} = \frac{(x_1 x_2 + y_1 y_2) + i(x_2 y_1 - x_1 y_2)}{x_2^2 + y_2^2}$$

i.e.

$$\frac{z_1}{z_2} = \frac{x_1 x_2 + y_1 y_2}{x_2^2 + y_2^2} + i\,\frac{x_2 y_1 - x_1 y_2}{x_2^2 + y_2^2} \tag{5.31}$$

and once more we see that the result is a complex number.

We thus see an important property of complex numbers, namely, that any arithmetic operation performed on two complex numbers produces another complex number.

Example 5.5   If $z_1 = -2 + 5i$ and $z_2 = 3 - i$ find

(a) $z_1 + z_2$ (b) $z_1 - z_2$ (c) $z_1 z_2$
(d) z_1/z_2 (e) z_1^2 (f) z_2^2

(a) Using (5.29), we have directly

$$z_1 + z_2 = -2 + 3 + i\{5 + (-1)\} = 1 + 4i$$

(b) Again using (5.29):

$$z_1 - z_2 = -2 - 3 + i\{5 - (-1)\} = -5 + 6i$$

(c) Applying (5.30), we have

$$z_1 z_2 = (-2)(3) - (5)(-1) + i\{(-2)(-1) + (3)(5)\}$$
$$= -6 + 5 + i(2 + 15) = -1 + 17i$$

(d) Substitution in (5.31) gives

$$\frac{z_1}{z_2} = \frac{(-2)(3) + (5)(-1)}{9 + 1} + i\,\frac{(3)(5) - (-2)(-1)}{9 + 1}$$

$$= \frac{-6 - 5}{10} + i\,\frac{15 - 2}{10} = -1.1 + 1.3i$$

(e) Using (5.30), we have

$$z_1^2 = z_1 z_1 = 4 - 25 + i(-10 - 10) = -21 - 20i$$

(f) Again using (5.30):

$$z_2^2 = z_2 z_2 = 9 - 1 + i(-3 - 3) = 8 - 6i$$

Hence there does not appear to be a simple relationship between a complex number and its square, but see equation (5.33) below.

The polar co-ordinate form of a complex number

Complex numbers in polar co-ordinate form have already been alluded to through the idea of modulus and argument (page 164). In transposing from Cartesian to polar co-ordinates we have the relationships $x = r\cos\theta$, $y = r\sin\theta$; so the complex number $z = x + iy$ is written in polar form as

$$z = r(\cos\theta + i\sin\theta) \qquad (5.32)$$

The multiplication of two complex numbers gives a particularly simple result in polar form. Let $z_1 = r_1(\cos\theta_1 + i\sin\theta_1)$ and $z_2 = r_2(\cos\theta_2 + i\sin\theta_2)$; then $z_1 z_2 = r_1 r_2(\cos\theta_1\cos\theta_2 + i\sin\theta_1\cos\theta_2 + i\cos\theta_1\sin\theta_2 - \sin\theta_1\sin\theta_2)$ and so

$$z_1 z_2 = r_1 r_2\{\cos(\theta_1 + \theta_2) + i\sin(\theta_1 + \theta_2)\} \qquad (5.33)$$

This last result follows from (4.7) and (4.6).

A MISCELLANY

Example 5.6 Express the numbers $z_1 = -2 + 5i$ and $z_2 = 3 - i$ in polar form, and hence form the product $z_1 z_2$.

The modulus of z_1 is

$$r = \sqrt{\{(-2)^2 + (5)^2\}} = \sqrt{(4 + 25)} = \sqrt{29}$$

The argument of z_1 is

$$\theta = \tan^{-1}(5/-2) = \tan^{-1}(-2.5)$$

So the polar form of z_1 is

$$z_1 = \sqrt{29}[\cos\{\tan^{-1}(-2.5)\} + i \sin\{\tan^{-1}(-2.5)\}]$$

Because approximations are involved in evaluating the terms in the expression, they are best left in the exact, albeit cumbersome, form as shown. Similarly for z_2 we have

$$z_2 = \sqrt{10}[\cos\{\tan^{-1}(-0.\dot{3})\} + i \sin\{\tan^{-1}(-0.\dot{3})\}]$$

Hence

$$z_1 z_2 = \sqrt{290}[\cos\{\tan^{-1}(-2.5) + \tan^{-1}(-0.\dot{3})\} + i \sin\{\tan^{-1}(-2.5) + \tan^{-1}(-0.\dot{3})\}]$$

A scientific calculator can now be employed to evaluate this expression; using a 10-digit calculator, we have

$$z_1 z_2 = 17.029\ 386\ 3\{\cos(-1.190\ 289\ 9 - 0.321\ 750\ 55) + i \sin(-1.190\ 289\ 9 - 0.321\ 750\ 55)\}$$

$$= 17.029\ 386\ 3\{\cos(-1.512\ 040\ 45) + i \sin(-1.512\ 040\ 45)\}$$

$$= 17.029\ 386\ 3(0.058\ 722\ 077 - 0.998\ 274\ 38i)$$

So

$$z_1 z_2 = 1.000\ 000\ 933 - 17.000\ 000\ 05i$$

which, apart from rounding error at the ends of the numbers and change of sign in both terms, is the same as the result of the corresponding multiplication in Example 5.5(c).

De Moivre's theorem

In equation (5.33), let $r_1 = r_2 = 1$, $\theta_1 = \theta_2 = \theta$, and so $z_1 = z_2 = z$. Then

$$z^2 = \cos 2\theta + i \sin 2\theta$$

Further, it can be shown that

$$(\cos \theta + i \sin \theta)^3 = \cos 3\theta + i \sin 3\theta$$

(see Exercise 5.6 at end of chapter), and indeed that for any real number n,

$$(\cos \theta + i \sin \theta)^n = \cos n\theta + i \sin n\theta \qquad (5.34)$$

The result embodied in (5.34) is known as **de Moivre's theorem**, and it leads to some very important relationships with wide practical applications.

The exponential form of a complex number

For the complex number of unit modulus

$$z = \cos \theta + i \sin \theta$$

we can differentiate with respect to θ:

$$\frac{dz}{d\theta} = -\sin \theta + i \cos \theta$$

$$= i^2 \sin \theta + i \cos \theta$$

since $i^2 = -1$

$$= i(\cos \theta + i \sin \theta)$$

So

$$\frac{dz}{d\theta} = iz \qquad (5.35)$$

From (5.35) we have

$$\frac{dz}{z} = i \, d\theta$$

so

$$\log_e z = i\theta + c$$

or

$$z = ae^{i\theta} \qquad \text{where } a = e^c$$

Now when $\theta = 0$, $z = 1$; so $a = 1$ and thus

$$z = e^{i\theta} \qquad (5.36)$$

and

$$e^{i\theta} = \cos \theta + i \sin \theta \qquad (5.37)$$

Combining the results of (5.33) and (5.37), again with $r = 1$, we have

$$e^{i\theta_1}e^{i\theta_2} = e^{i(\theta_1 + \theta_2)} \qquad (5.38)$$

From de Moivre's theorem, equation (5.34),

$$(e^{i\theta})^n = e^{in\theta} \qquad (5.39)$$

169

Again, from (5.34) we have

$$\{\cos(\theta/n) + i \sin(\theta/n)\}^n = \cos \theta + i \sin \theta$$

Taking the nth root of both sides gives

$$\cos(\theta/n) + i \sin(\theta/n) = (\cos \theta + i \sin \theta)^{1/n}$$

or

$$e^{i\theta/n} = (e^{i\theta})^{1/n} \tag{5.40}$$

Finally, using (5.39) to take the mth power of (5.40) gives

$$e^{im\,\theta/n} = (e^{i\theta})^{m/n} \tag{5.41}$$

Relationships (5.38) to (5.41) show that exponential forms of two complex numbers obey all the normal laws of indices. It can also be shown (Exercise 5.7) that

$$e^{-i\theta} = \cos \theta - i \sin \theta \tag{5.42}$$

Partial fractions

Partial fractions were introduced in Causton (1983) as a method of integration, and a few examples of the resolution of some simple fractions into partial fractions were given. Partial fractions have other uses, notably in the Laplace transform method of solving linear differential equations, and more systematic rules and illustrative examples will now be given.

Our starting point is the fraction $p(x)/q(x)$ where $p(x)$ and $q(x)$ are polynomial functions of x. Further, the fraction must be *proper*, that is, the degree of $p(x)$ is less than that of $q(x)$; if not, an algebraic long division must be done first (details can be found in a textbook of elementary algebra). Next, $q(x)$ must be factorised into irreducible factors, i.e. it is expressed as the product of linear factors $(x + a)$ and/or irreducible quadratic factors $(x^2 + bx + c)$†. Each factor will then be the denominator of a partial fraction, which has a numerator of one degree lower than the denominator. The determination of the numerator of each partial fraction differs slightly according to the nature of the denominator.

Denominator of linear factors only

This is the simplest case, and we expect that the numerator of each partial fraction will be a constant, i.e.

$$\frac{p(x)}{q(x)} \equiv \frac{p(x)}{(x + a)(x + b)} \equiv \frac{A}{x + a} + \frac{B}{x + b}$$

†An irreducible quadratic expression is one whose linear factors are complex numbers.

Example 5.7   Resolve the following into partial fractions:

(a) $\dfrac{x^2 - 5x + 2}{(x-1)(x-2)(x-3)}$ (b) $\dfrac{2+x}{x-x^2}$

(a)   Assume that

$$\frac{x^2 - 5x + 2}{(x-1)(x-2)(x-3)} \equiv \frac{A}{x-1} + \frac{B}{x-2} + \frac{C}{x-3}$$

Multiply both sides by the denominator of the left-hand side, giving

$$x^2 - 5x + 2 = A(x-2)(x-3) + B(x-1)(x-3) + C(x-1)(x-2)$$

$$(5.43)$$

Put $x = 1$ in (5.43); then

$1 - 5 + 2 = A(-1)(-2)$,   i.e. $2A = -2$ or $A = -1$

Put $x = 2$ in (5.43), then

$4 - 10 + 2 = B(1)(-1)$,   i.e. $-B = -4$ or $B = 4$

Put $x = 3$ in (5.43), then

$9 - 15 + 2 = C(2)(1)$,   i.e. $2C = -4$ or $C = -2$

Hence

$$\frac{x^2 - 5x + 2}{(x-1)(x-2)(x-3)} \equiv \frac{-1}{x-1} + \frac{4}{x-2} - \frac{2}{x-3}$$

(b)   The denominator is not an irreducible quadratic because it can be factorised into two linear factors $x(1 - x)$. So

$$\frac{2+x}{x-x^2} \equiv \frac{2+x}{x(1-x)} \equiv \frac{A}{x} + \frac{B}{1-x} \qquad (5.44)$$

Multiply middle and right-hand sides of (5.44) by $x(1 - x)$, which gives

$$2 + x = A(1-x) + Bx \qquad (5.45)$$

Put $x = 0$ in (5.45), then $A = 2$. Put $x = 1$ in (5.45), then $B = 3$. So

$$\frac{2+x}{x-x^2} \equiv \frac{2+x}{x(1-x)} \equiv \frac{2}{x} + \frac{3}{1-x}$$

Example 5.7(b) shows that sometimes a slight adjustment has to be made to the denominator of a fraction in order to obtain irreducible terms, either linear or quadratic. A quadratic function will of course always have two

171

linear factors, but these may be complex; an irreducible quadratic is one which has complex linear factors. The following examples, which all have the same numerator (for comparative purposes), show the possibilities.

Example 5.8   Resolve the following into partial fractions:

(a) $\dfrac{4x}{x^2 - 5x + 6}$ (b) $\dfrac{4x}{x^2 + x - 1}$

(c) $\dfrac{4x}{x^2 + x + 1}$ (d) $\dfrac{4x}{(x + 1)(x^2 - 1)}$

(a)   The denominator here has obvious factors: $(x - 3)(x - 2)$ (two numbers which add to give -5 and multiply to give 6). Thus we may write

$$\frac{4x}{x^2 - 5x + 6} \equiv \frac{4x}{(x - 3)(x - 2)} \equiv \frac{A}{x - 3} + \frac{B}{x - 2}$$

Multiply the middle and right-hand sides by $(x - 3)(x - 2)$:

$$4x \equiv A(x - 2) + B(x - 3)$$

$$\text{put } x = 3, \qquad \text{then } A = 12$$

$$\text{put } x = 2, \qquad \text{then } B = -8$$

So

$$\frac{4x}{x^2 - 5x + 6} \equiv \frac{4x}{(x - 3)(x - 2)} \equiv \frac{12}{x - 3} - \frac{8}{x - 2}$$

(b)   Here the denominator does not have obvious factors. The easiest way to proceed is to treat the denominator as a quadratic equation, and use the standard formula for its solution. So, for the equation

$$x^2 + x - 1 = 0$$

we have

$$x = \frac{-1 \pm \sqrt{(1 + 4)}}{2}$$

So either

$$x = -\tfrac{1}{2} + \frac{\sqrt{5}}{2} = 0.6180$$

or

$$x = -\tfrac{1}{2} - \frac{\sqrt{5}}{2} = -1.6180$$

(both results to four decimal places). Hence

$$x^2 + x - 1 = (x - 0.6180)(x + 1.6180)$$

Thus, we can write

$$\frac{4x}{x^2 + x - 1} \equiv \frac{4x}{(x - 0.618)(x + 1.618)} \equiv \frac{A}{x - 0.618} + \frac{B}{x + 1.618}$$

Proceeding as before, we have

$$4x \equiv A(x + 1.618) + B(x - 0.618)$$

put $x = 0.618$,   then $2.472 = 2.236A$,   i.e. $A = 1.1056$

put $x = -1.618$,   then $-6.472 = -2.236B$,   i.e. $B = 2.8945$

Hence

$$\frac{4x}{x^2 + x - 1} \equiv \frac{4x}{(x - 0.618)(x + 1.618)} = \frac{1.1056}{x - 0.618} + \frac{2.8945}{x + 1.618}$$

$$(5.46)$$

If the right-hand side of (5.46) is reamalgamated as a check, an error will be found in the 4th decimal place since A and B were only evaluated to this degree of accuracy.

(c)   The quadratic denominator is irreducible and so the expression cannot be further simplified.

(d)   At first sight, one might tend to regard the denominator as the product of a linear factor and an irreducible quadratic. However, if you proceed on this basis, you will find the present method will fail. In fact, the quadratic term is not irreducible because $(x^2 - 1)$ *can* be factorised to give $(x + 1)(x - 1)$. So the denominator can be written as $(x + 1)(x + 1)(x - 1)$, i.e. as $(x + 1)^2(x - 1)$, and we have

$$\frac{4x}{(x + 1)(x^2 - 1)} \equiv \frac{4x}{(x + 1)^2(x - 1)}$$

This involves what is known as a **repeated linear factor** which is the subject of the next section, so the continuation of this example will appear as Example 5.9(a).

Denominator containing repeated linear factors

Suppose in the fraction $p(x)/q(x)$, $q(x) = (x + a)(x + b)^n$, where n is a positive integer. To write partial fractions as $A/(x + a) + B/(x + b)^n$ will lead to failure; we need several more partial fractions where the power of

A MISCELLANY

the linear factor $(x+b)$ is 'built-up' successively one at a time, that is, we must write

$$\frac{p(x)}{(x+a)(x+b)^n} \equiv \frac{A}{x+a} + \frac{B}{x+b} + \frac{C}{(x+b)^2} + \ldots + \frac{Z}{(x+b)^n} \qquad (5.47)$$

Example 5.9 *Resolve the following into partial fractions:*

(a) $\dfrac{4x}{(x+1)^2(x-1)}$ (b) $\dfrac{16-88x+89x^2-13x^3}{x^3(x-4)^2}$

(a) From (5.47), we have

$$\frac{4x}{(x+1)^2(x-1)} \equiv \frac{A}{x+1} + \frac{B}{(x+1)^2} + \frac{C}{x-1}$$

i.e.

$$4x \equiv A(x+1)(x-1) + B(x-1) + C(x+1)^2$$

put $x=1$,   then $4=4C$,   i.e. $C=1$

put $x=-1$,   then $-4=-2B$,   i.e. $B=2$

Now put $x=0$, and use the results for B and C already obtained; this gives

$$0 = -A - 2 + 1, \quad \text{i.e. } A = -1$$

So

$$\frac{4x}{(x+1)^2(x-1)} \equiv -\frac{1}{x+1} + \frac{2}{(x+1)^2} + \frac{1}{x-1}$$

(b) Here we must write

$$\frac{16-88x+89x^2-13x^3}{x^3(x-4)^2} \equiv \frac{A}{x} + \frac{B}{x^2} + \frac{C}{x^3} + \frac{D}{x-4} + \frac{E}{(x-4)^2}$$

i.e.

$$16-88x+89x^2-13x^3 \equiv Ax^2(x-4)^2 + Bx(x-4)^2 + C(x-4)^2 + Dx^3(x-4) + Ex^3$$

put $x=0$,   then $16=16C$,   i.e. $C=1$

put $x=4$,   then $16-352+1424-832=64E$,   i.e. $E=4$

By putting x equal to any other three convenient values successively, and using the values of C and E already obtained, we set up three linear equations to solve for the remaining constants A, B, D:

put $x = 1$,   then $16 - 88 + 89 - 13$

$$= A(1)(9) + B(1)(9) + 9 + D(1)(-3) + 4$$

i.e. $9A + 9B - 3D = -9$

put $x = 2$,   then $16 - 176 + 356 - 104$

$$= A(4)(4) + B(2)(4) + 4 + D(8)(-2) + 32$$

i.e. $16A + 8B - 16D = 56$

put $x = 3$,   then $16 - 264 + 801 - 351$

$$= A(9)(1) + B(3)(1) + 1 + D(27)(-1) + 108$$

i.e. $9A + 3B - 27D = 93$

Dividing the first and third of the above equations throughout by 3, and the second by 8, we have the set

$$3A + 3B - D = -3$$

$$2A + B - 2D = 7$$

$$3A + B - 9D = 31$$

By one of the methods given in Chapter 1, we may compute that $A = 3$, $B = -5$, $D = -3$; hence

$$\frac{16 - 88x + 89x^2 - 13x^3}{x^3(x-4)^2} \equiv \frac{3}{x} - \frac{5}{x^2} + \frac{1}{x^3} - \frac{3}{x-4} + \frac{4}{(x-4)^2}$$

Denominator containing irreducible quadratic factors

We now suppose that in the fraction $p(x)/q(x)$, $q(x) = (x + a)(x^2 + bx + c)$ where $(x^2 + bx + c)$ is irreducible; the appropriate form of partial fractions is now

$$\frac{p(x)}{(x + a)(x^2 + bx + c)} \equiv \frac{A}{x + a} + \frac{Bx + C}{x^2 + bx + c}$$

The numerator of the fraction containing the quadratic denominator is not a constant but a linear function of x.

Example 5.10   *Resolve the following into partial fractions:*

(a) $\dfrac{3(x + 3)}{(x + 1)(x^2 + x + 1)}$ (b) $\dfrac{1 - 2x}{x^3 + 1}$

175

(a) The quadratic term is irreducible, so we write

$$\frac{3(x+3)}{(x-1)(x^2+x+1)} \equiv \frac{A}{x-1} + \frac{Bx+C}{x^2+x+1}$$

i.e.

$$3(x+3) \equiv A(x^2+x+1) + (Bx+C)(x-1)$$

put $x = 1$,   then $12 = 3A$,   i.e. $A = 4$

put $x = 0$,   then $9 = 4 - C$,   i.e. $C = -5$

put $x = 2$,   then $15 = (4)(7) + (2B - 5)$

       i.e. $-8 = 2B$,   and so $B = -4$

Hence

$$\frac{3(x+3)}{(x-1)(x^2+x+1)} \equiv \frac{4}{x-1} - \frac{4x+5}{x^2+x+1}$$

(b) The denominator must be factorised, and we find that $x^3 + 1 \equiv (x+1)(x^2-x+1)$. The quadratic term is irreducible so we write our problem as

$$\frac{1-2x}{x^3+1} \equiv \frac{1-2x}{(x+1)(x^2-x+1)} \equiv \frac{A}{x+1} + \frac{Bx+C}{x^2-x+1}$$

i.e.

$$1-2x \equiv A(x^2-x+1) + (Bx+C)(x+1)$$

put $x = -1$,   then $3 = 3A$,   i.e. $A = 1$

put $x = 0$,   then $1 = 1 + C$,   i.e. $C = 0$

put $x = 1$,   then $-1 = 1 + 2B$,   i.e. $B = -1$

So

$$\frac{1-2x}{x^3+1} \equiv \frac{1-2x}{(x+1)(x^2-x+1)} \equiv \frac{1}{x+1} - \frac{x}{x^2-x+1}$$

This example shows that, in effect, we introduced too many constants since $C = 0$; but we only know this through hindsight. Always in partial fraction work one must adhere to the rules.

Numerical analysis

Numerical analysis – or **numerical methods** might be the better name – is nowadays a subject of considerable importance to the *user* of mathematics why?

The non-mathematician who actually uses mathematics in his or her field of study to analyse a particular problem normally requires numerical answers. Such a person may or may not be interested in the intervening mathematical steps which eventually produce a formula or equation from which the numerical results are obtained, but on the other hand the original formulation of the problem in quantitative terms requires an intelligent application of mathematical principles – hence the purpose of this and many other books written about mathematical methods for the non-mathematician. But however desirable a good knowledge of the underlying mathematics we are using is, a result in terms of numbers is the desired outcome; and it is often found in practice that to obtain a numerical result from a mathematical formulation is far from straightforward.

Take an ordinary quadratic equation, for example

$$2x^2 - 7x + 4 = 0$$

We know that x may be evaluated (i.e. the equation can be solved; or, to put it another way, its roots may be found) by a formula (Causton 1983, eqn (4.2)). An alternative way of expressing this is to say that an **analytical solution** exists for the equation. But even here things are not entirely straightforward as the equation contains a square root, and it is very rare in practice for the square root to yield a rational number which can be written down precisely. In a textbook, where the aim is to discuss quadratic equations themselves and how to obtain their roots, examples are usually constructed to give rational answers; but in the real world, roots of quadratic equations are almost invariably irrational. Immediately we have the problem that the answer is accurate only to a certain number of decimal places. If the formula has been personally evaluated on a calculator, we may have confidence in asserting that the answer is correct to a certain number of decimal places; if, however, a prewritten program on a micro or mainframe computer has been used, then one cannot be so sure as to the degree of accuracy of the answer because there is less personal involvement with the computational steps of the actual problem in hand.

If the numerical solution of our quadratic equation gives us the final answers required, the above difficulty is not serious; but if the answers are only an intermediate step in a long series of processes in a computer program, the situation may be different. Thus if several numerical results are produced in stages, each with a degree of error, the errors may accumulate to such a degree that the final answers are useless, even nonsensical. We shall return to this part of the discussion later.

Now consider a cubic equation, e.g.

$$x^3 - 2x^2 - 4x + 1 = 0$$

Methods of obtaining an analytical solution are available, but they are more difficult to use than in the case of a quadratic equation where one merely 'plugs' the coefficient values into a formula. A similar situation holds for the quartic equation, but no analytical solutions exist for polynomial equations of higher degree. However, almost any kind of equation can be solved by numerical methods.

In a different context, you already know that some functions cannot be integrated; in other words, in terms of the notation

$$\int f(x) = F(x) + c \qquad (5.48)$$

there are forms of $f(x)$ for which no known method will enable us to obtain $F(x)$. But for

$$\int_a^b f(x) = F(b) - F(a) \qquad (5.49)$$

which gives a numerical result, numerical methods (e.g. Simpson's rule) will give us at least an approximate solution almost regardless of what the functional form, $f(x)$, may be. The process of obtaining $F(x)$ from $f(x)$ in (5.48) which, of course, is part of the process of evaluating the definite integral, (5.49), may be called the *analytical* evaluation of (5.49), as opposed to a *numerical* method such as Simpson's rule.

Yet another example of the application of numerical methods will be met with in Chapter 7 on differential equations. Many of the differential equations arising in practice do not have analytical solutions, and recourse must be had to numerical ones.

The impression you must have gained by now is that numerical methods are applied when analytical ones fail. Yet the bulk of mathematics books, and this one is no exception, concentrate on analytical methods. There are three reasons for this:

(1) analytical methods of solution arise naturally in the presentation of the principles and methods of a mathematical topic;
(2) analytical solutions, if obtainable, are better than numerical ones; and
(3) analytical solutions provide the necessary 'test data' to check that a numerical method is working properly.

The word 'better' in (2) above means different things according to context. In the solution of ordinary equations and in the evaluation of definite integrals analytical solutions are usually, but not always, more accurate. In the field of differential equations, the analytical solution is another equation without derivatives. The analytical solution, which provides the solution in an explicit functional form, is greatly preferable to the numerical solution which can only be set out as a table of values or as a graph over a certain range.

The subject of numerical analysis is now very large, and it is outside the scope of this book to cover the subject in detail; some suitable books are listed at the end of this chapter. However, some reference to numerical methods inevitably arises in books which primarily concentrate on mathematical methods. Thus Simpson's rule for numerical integration, or numerical quadrature as it is commonly called, has been described in Causton (1983); a simple numerical approach to the solution of a first-order ordinary differential equation will be discussed in Chapter 7. It is appropriate here to present a numerical method of solving an ordinary equation.

An iterative method for estimating the roots of a function

An iterative numerical procedure is essentially one of organised trial and error. One starts with a guess at the answer, then the iterative procedure should provide a value that is closer to the true value than was the original guess; this new value is again used as 'input' to the procedure which should then give a result even closer to the correct one. In theory, we may iterate as many times as desired using the result of the previous iteration as input to the current one and, if all goes well, we should converge on the answer. Another way of saying this is that we can obtain the answer to any desired level of accuracy by performing a sufficient number of iterations. However, in practice, things rarely go as smoothly as this!

Probably the best known numerical technique for obtaining the roots of an equation is the **Newton–Raphson method**, which will be described using a third-degree polynomial as an example. First, however, we need to examine the nature of the roots of the cubic polynomial; this is an analytical process.

THE ROOTS OF THE THIRD-DEGREE POLYNOMIAL

The real roots of the equation

$$a + bx + cx^2 + dx^3 = 0 \qquad (5.50)$$

specify the point or points of intersection of the curve with the x-axis. In any cubic curve there will always be one such point, and there may be three if the curve contains a loop which crosses the x-axis. The intermediate situation of two real roots (i.e. one distinct and one repeated real root) is that in which the bottom or top of the loop, as the case may be, just touches the x-axis. To ascertain the number of real roots, the **discriminant** (Causton 1983, p. 46) of the equation has to be evaluated; the following paragraph describes how this is done.

First, the equation has to be obtained in a so-called standard form; in this form there is no term in the square of the variable, and the coefficient of the cubic term is unity. Putting the equation into standard form involves

179

defining a new variable, z, which is formed by adding $c/3d$ to the original variable, x, and so $z = x + c/3d$ or $x = z - c/3d$. Then we substitute for x in the original equation

$$a + b\left(z - \frac{c}{3d}\right) + c\left(z - \frac{c}{3d}\right)^2 + d\left(z - \frac{c}{3d}\right)^3 = 0$$

From this equation we obtain, after some algebra

$$\frac{1}{d^3}(ad^2 - bcd + c^3d - c^3) + \frac{1}{d^2}(bd - c^3)z + z^3 = 0 \qquad (5.51)$$

Notice the absence of a term in z^2, and that the coefficient of z^3 is 1. Now let $p = (bd - c^2)/d^2$ and $q = (ad^2 - bcd + c^3d - c^3)/d^3$; then (5.51) can be written as

$$z^3 + pz + q = 0 \qquad (5.52)$$

Equation (5.52) is the standard cubic equation, and the quantity $p^3 + 27q^2$ is the discriminant. If the discriminant is negative, (5.52) has three real roots; if it is positive, there is only one real root; and if it is zero, we have the intermediate condition of two real roots, i.e. two coincident roots and one other.

In obtaining the standard form of the equation, we obtained a new variable, z, merely by adding a constant, $c/3d$, to the original variable, x.

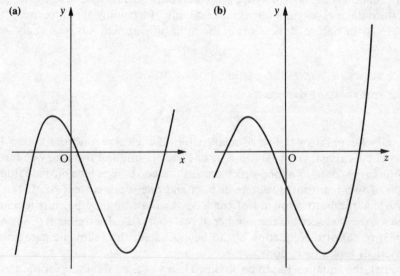

Figure 5.4   (a) and (b) Two third-degree polynomials differing only in their positions on the x-axis.

This means that the cubic in z, (5.51), has a curve of exactly the same shape and bears exactly the same relationship to the y-axis as does the original equation, (5.50), there is merely a displacement along the horizontal axis and no displacement vertically. Hence the number of roots established by the discriminant for the standard form of the equation will be the same as for the original. The situation is shown in Figure 5.4.

Example 5.11   *How many roots has the cubic equation*

$$x^3 - 2x^2 - 4x + 1 = 0? \tag{5.53}$$

Matching the coefficients with the symbolic coefficients of (5.50), $a = 1$, $b = -4$, $c = -2$, $d = 1$. Thus $c/3d = -2/3$, and so $z = x + (-2/3) = x - 2/3$ and $x = z + 2/3$. Substituting for x in (5.53) gives

$$(z + \tfrac{2}{3})^3 - 2(z + \tfrac{2}{3})^2 - 4(z + \tfrac{2}{3}) + 1 = 0$$

After multiplying out the brackets and collecting together like terms, we have

$$z^3 - \frac{16}{3} z - \frac{61}{27} = 0 \tag{5.54}$$

Equation (5.54) is the standard form of (5.53). Next, compare the coefficients of (5.54) with those in the general standard cubic, equation (5.52); $p = -16/3$, $q = -61/27$. The discriminant is $p^3 + 27q^2 = -4096/27 + 3721/27 = -375/27$. Therefore, as the discriminant is negative, equation (5.54), and so equation (5.53) has three real roots.

THE NEWTON–RAPHSON METHOD

Figure 5.5 shows a graph of the function $y = f(x)$. The curve intersects the x-axis at the point P, distant X from the origin, and this represents a root of the function (there may or may not be others); so that when $x = X$, then $f(x) = 0$, i.e. $f(X) = 0$. The problem is to find the value of X.

Assume that a guess of the value can be made, X_0, represented on the diagram as point Q. Erect a vertical line from Q to intersect the curve of $f(x)$ at R (the length of QR is Y_0, since R lies on the curve of $y = f(x)$); and, finally, draw in the tangent to the curve at R, extending it to intersect the x-axis at S. Let point S be distant X_1 from the origin. Now in triangle QRS by elementary trigonometry

$$\tan \theta = \frac{QR}{QS} = \frac{Y_0}{X_0 - X_1} \tag{5.55}$$

Now $Y_0 = f(X_0)$ and $\tan \theta = f'(X_0)$. The latter is because $\tan \theta = f'(x)$ at the point where $x = X_0$. Hence, from the extreme left- and right-hand sides of

181

(5.55), we have

$$f'(X_0) = \frac{f(X_0)}{X_0 - X_1}$$

and rearrangement gives

$$X_1 = X_0 - \frac{f(X_0)}{f'(X_0)} \qquad (5.56)$$

In Figure 5.5 it is evident that point S, where $x = X_1$, is nearer to point P, where $x = X$ (the actual root being sought), than is the initial guess of the root, $x = X_0$, represented by point Q. Relationship (5.56) shows that X_1, which is a better estimate of the actual root ($x = X$) than is the initial guess ($x = X_0$), can be obtained from X_0 by dividing the value of the function at X_0 by the value of the first derivative at X_0 and subtracting the quotient so obtained from X_0. In general terms, X_0 is called a starting value, X_1 the adjusted value, and the process of obtaining X_1 from X_0 an iteration. A starting value has to be provided before the Newton–Raphson process can be used. Occasionally a starting value can be provided by some other mathematical method, but a more generally useful procedure is to sketch a graph of the function. The iterative process, that is the procedure summarized in (5.56), may be repeated as often as desired using the adjusted value of the previous iteration as the starting value of each new iteration. It will usually be observed that the difference between starting and adjusted values decreases with successive iterations, so that although the required root cannot be obtained *exactly* by this procedure, it can be approximated to any desired degree of accuracy.

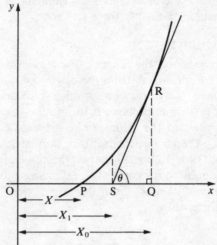

Figure 5.5 The Newton–Raphson method for approximating the root of a function.

Example 5.12   Find the roots of the function

$$f(x) = x^3 - 2x^2 - 4x + 1.$$

Before the Newton Raphson technique can be used, starting values must be provided, and we shall sketch the curve of the function to assist in this task.

The phrase 'sketch the curve', as opposed to the expression 'draw a graph of the curve', implies that we only require a rough sketch of the curve to show its main features. First, some points on the curve must be calculated, and this is conveniently done as shown in the table below.

x	x^2	x^3	$-2x^2$	$-4x$	$f(x) = x^3 - 2x^2 - 4x + 1$
3	9	27	-18	-12	-2
4	16	64	-32	-16	17
-2	4	-8	-8	8	-7

A maximum point $(-\frac{2}{3}, 2\frac{13}{27})$, a minimum $(2, -7)$, and a point of inflexion $(\frac{2}{3}, -2\frac{7}{27})$ are also obtainable by the usual methods, and these are plotted onto the sketch first. Also, inspection of the equation shows that the intercept is 1, i.e. the point $(0, 1)$ lies on the curve. Through these four points, sketch the curve from the maximum to the minimum. Evidently there is one root near $x = 0$.

Now consider the right-hand side of the graph. Calculation of points on the curve at $x = 3$ and $x = 4$, followed by sketching the curve in this region, shows there to be a root near $x = 3$; and finally, on the left-hand side of the graph, we find a root near $x = -1$. The completed sketch is shown in Figure 5.6.

In employing (5.56) to estimate the roots, note that

$$f(x) = x^3 - 2x^2 - 4x + 1$$

and

$$f'(x) = 3x^2 - 4x - 4$$

The use of the Newton–Raphson technique will be demonstrated in finding the middle root of the equation, near $x = 0$, and this figure will be used as the starting value. So we have $X_0 = 0$, $f(0) = 1$, and $f'(0) = -4$. Substituting in (5.56):

$$X_1 = 0 - \frac{1}{-4} = \frac{1}{4} = 0.25$$

This is one iteration, in which the initial 'guess' of the root, $x = 0$, has been 'improved' to $x = 0.25$. Now use this value of x as the starting value for a

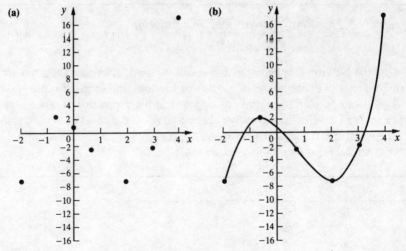

Figure 5.6 (a) Points on the curve of the function $f(x) = x^3 - 2x^2 - 4x + 1$. (b) The same, but with curve drawn through the points.

second iteration: $X_1 = 0.25$, $f(0.25) = 0.015\,625 - 0.1250 - 1 + 1 = -0.109\,375$, and $f'(0.25) = 0.1875 - 1 - 4 = -4.8125$. Substituting in (5.56), we have

$$X_2 = 0.25 - \frac{-0.109\,375}{-4.8125} = 0.25 - 0.0227 = 0.227\,273.$$

For a further iteration we have $X_2 = 0.227\,273$, $f(0.227\,273) = 0.011\,739 - 0.103\,306 - 0.909\,092 + 1 = -0.000\,659$ and $f'(0.227\,273) = 0.1549\,59 - 0.909\,092 - 4 = -4.754\,133$. So

$$X_3 = 0.227\,273 - \frac{-0.000\,659}{-4.754\,133} = 0.227\,273 - 0.000\,139 = 0.227\,134$$

You will notice that the first three places of decimals have remained unchanged between X_2 and X_3; thus one of the roots of the function $f(x) = x^3 - 2x^2 - 4x + 1$ is 0.227 (correct to three decimal places).

Suggestions for further reading

Most of the topics covered in this chapter are standard items in mathematic textbooks, the exception being numerical analysis.

A brief, terse introduction to this subject is given by **Wilkes, M. V.** (1966). *A shor introduction to numerical analysis*. Cambridge: Cambridge University Press.

A longer, much more readable but yet rigorous, book is **Acton, F. S. (1970).** *Numerical methods that work.* New York: Harper & Row. Although specifically written for students of engineering, the methods described are of general applicability; and the style is exactly right for the non-mathematician.

Since numerical analysis has been revolutionised by electronic computation in all its forms, it would in general be unwise to study books on this subject written much before about 1970.

Exercises

5.1 Expand, up to the fourth term:

    (a) $1/(1 - x)$ (b) $(4 + x^2)$ (c) $\sqrt[3]{(8 - x^3)^2}$

5.2 Using the Maclaurin series, show that

$$\cos x = 1 - \frac{x^2}{2!} + \frac{x^4}{4!} - \frac{x^6}{6!} + \ldots$$

5.3 By using the appropriate series, evaluate

    (a) $\sin 50°$ (b) $\cos 20°$ (c) $\log_e 2$ (d) $\log_e 0.8$

5.4 Write down the modulus and argument of each of the following complex numbers, and plot them on an Argand diagram:

    (a) $z_1 = 4 - 2i$ (b) $z_2 = -5 - i\sqrt{3}$
    (c) $z_3 = 4 + 2i$ (d) $z_4 = -5 + i\sqrt{3}$

5.5 Perform the following arithmetic operations on the complex numbers defined in 5.4 above:

    (a) $z_1 + z_2$ (b) $z_1 + z_3$ (c) $z_1 - z_3$ (d) $z_2 + z_4$
    (e) $z_2 - z_4$ (f) $z_1 z_4$ (g) $z_1 z_3$ (h) $z_2 z_4$
    (i) z_2/z_3 (j) z_1/z_3 (k) z_2/z_4

5.6 Prove that $(\cos \theta + i \sin \theta)^3 = \cos 3\theta + i \sin 3\theta$.

5.7 Prove that $e^{-i\theta} = \cos \theta - i \sin \theta$.

5.8 Express the following in partial fractions

    (a) $\dfrac{4x + 7}{(x - 2)(2x + 1)}$ (b) $\dfrac{3x + 1}{(x + 2)^2}$ (c) $\dfrac{5x - 7}{(x + 3)(x^2 + 2)}$

5.9 Determine the number of real roots of the cubic polynomial $x^3 - 3x + 3$, and use the Newton–Raphson technique to determine the root(s) correct to four decimal places.

6

Functions of more than one variable

So far in this book when considering mathematical functions, they have usually been functions of one variable only; in general we write $y = f(x)$, $f(x)$ being a function of the one variable x. Although there are very many useful applications of functions of a single variable, it would be frustratingly restrictive to have to confine ourselves to this case. We can quite easily think of many situations where a quantity depends on several others instead of just one other quantity:

(a) the growth rate of a population depends on the number of organisms present and the level of one overriding limiting factor in the environment;

(b) the concentration of a drug in the blood steam at one point in the body of an animal, which is being injected at another point in the body, depends on the rate of injection and the rate of metabolic breakdown or removal of the drug from the blood, together with the rate of blood flow along the vein;

(c) the growth rate of a plant depends on the availability of several mineral elements.

Example (a) is essentially a function of two variables, (b) is a function of three, and (c) is a function of an unspecified large number of variables.

The concept of a function of several variables mainly involves extensions of the features of a function of a single variable into higher dimensions – three dimensions for a function of two variables, four dimensions for a function of three variables, and so on; but there are properties of a function of two or more variables which have no analogues in the function of a single variable. Fortunately we are able to represent graphically a function of two variables in three-dimensional space, and functions of three or more variables, which cannot be depicted graphically, involve no further new ideas.

The first half of this chapter deals with the two simplest kinds of function of two variables only, namely, the linear function, which is an extension of the straight line, and the quadratic function, which is an extension to three dimensions of the second-degree polynomial. Both these functions of one variable are discussed in Causton (1983, Chs 3 & 4, respectively), and you are advised to read these sections before proceeding with the current material. Indeed Causton (1983, Ch. 3) does introduce the three-dimensional analogue of the straight line. There, the axes were designated x_1 and x_2 for the two independent variables and y for the dependent variable, giving, in general notation, $y = f(x_1, x_2)$, i.e. y is a function of the two variables x_1 and x_2. Denoting the independent variables by subscripted xs simplifies the extension of the notation to several independent variables. However, in the bulk of this chapter, in which we are dealing with two independent variables only, we shall denote them by x and y with z representing the dependent variable; hence $z = f(x, y)$. This is done for notational clarity.

Linear functions

The plane $z = a + bx + cy$

The three-dimensional analogue of the two-dimensional straight line is a plane surface, represented in Figure 6.1 as the figure IJKL with respect to co-ordinate axes x, y (dependent variables), and z (independent variable). The equation of this plane is

$$z = a + bx + cy \tag{6.1}$$

The intersection of the plane with the x–z axial plane is the straight line IJ, which has gradient b (negative in this example) and intercept on the z-axis of a. Hence the equation of line IJ is

$$z = a + bx \tag{6.2}$$

Similarly, the intersection of the plane with the y–z axial plane is the straight line IL, which has gradient c (positive in this example), and again the intercept on the z-axis is a. Thus the equation of line IL is

$$z = a + cy \tag{6.3}$$

The intersection of the plane with the x–y axial plane in the straight line JK is also important. Here $z = 0$, and so (6.1) becomes

$$a + bx + cy = 0$$

or

$$y = -\frac{a}{c} - \frac{b}{c} x \tag{6.4}$$

187

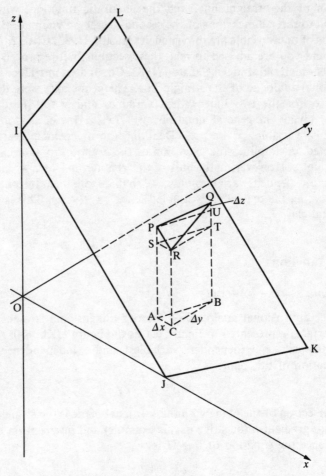

Figure 6.1 A plane surface showing increment details from a point on the surface.

The gradient of line JK is $-b/c$, which is positive in this example (see Fig. 6.1) because b is negative and c is positive, and the intercept on the y-axis is negative (KJ produced) because both a and c are positive.

GRADIENTS OF THE PLANE

A straight line has a constant gradient. Similarly a plane has a constant gradient *in a specified direction*. We have already found that the gradient of the plane in the x–z axial plane is b. Along the line IJ $y = 0$ (constant). For any other plane parallel to the x–z axial plane, y is also a constant and the gradient of the plane surface under consideration in the x–z direction is also b. To see this algebraically, let $y = Y$, a constant value. Substituting into (6.1) gives

$$z = a + bx + cY$$

but since both c and Y are constants the term cY is also a constant and can be amalgamated with a, giving

$$z = (a + cY) + bx$$

a straight line with gradient b.

Graphically (Fig. 6.1), let P be a point on the plane and R also a point on the plane distant Δx from P in the x-direction, with both P and R having the same y-value (see Causton 1983, p. 68, for details of this notation). Then PACR defines a vertical plane parallel to the x–z axial plane. The gradient of PR is b, and so PS/SR $= b$. Now SR $=$ AC $= \Delta x$, and so

$$PS = b \, \Delta x \tag{6.5}$$

Similar arguments apply to a gradient on the plane parallel to the y–z axial plane. If Q and R are points on the plane whose distance apart is Δy in the y-direction and which have the same x co-ordinate value, then QT/RT $= c$, and so

$$QT = c \, \Delta y \tag{6.6}$$

More generally, consider points PQ on the plane, distance apart Δx in the x-direction and Δy in the y-direction. What is QU ($= \Delta z$) and so the gradient of PQ relative to the x–y plane or line ST which is parallel to AB? The **total increment**, Δz, is going to be the sum of the increment in z due to Δx, which is PS, and the increment in z due to Δy, which is QT. Hence $\Delta z = $ PS $+$ QT, and from (6.5) and (6.6) we have

$$\Delta z = b \, \Delta x + c \, \Delta y \tag{6.7}$$

In the example shown in Figure 6.1, Δz is smaller than either of its components because b, the gradient in the x–z direction, is negative (see Example 6.1 below). Further, since AB $= \sqrt{\{(\Delta x)^2 + (\Delta y)^2\}}$, the gradient of PQ is given by

$$G = \frac{b \, \Delta x + c \, \Delta y}{\sqrt{\{(\Delta x)^2 + (\Delta y)^2\}}} \tag{6.8}$$

Of course, if the x and y co-ordinates of points P and Q are known, Δz can be calculated by direct substitution into (6.1). The result expressed in the form of (6.7) is of importance for later material (page 216).

189

Example 6.1   The equation of the plane surface shown in Figure 6.1 is

$$z = 21 - x + \tfrac{1}{2}y$$

and P and Q are two points on the plane whose x co-ordinates are 13 and 15, respectively, and whose y-co-ordinates are 7 and 13, respectively. Find:

(a)   *the equation of the line of intersection of the plane with the x–y axial plane;*

(b)   *the total increment in z between the points P and Q;*

(c)   *the gradient of PQ relative to the x–y axial plane.*

(a)   Equating the coefficients of our present equation with those of the general form (6.1), we have $a = 21$, $b = -1$, $c = \tfrac{1}{2}$. Hence, from (6.4), we have

$$y = -42 + 2x$$

So the gradient of JK in the $x–y$ plane is 2.

(b)   From the co-ordinates given, $\Delta x = 2$ and $\Delta y = 6$. Substituting into (6.7) gives

$$\Delta z = (-1)(2) + (\tfrac{1}{2})(6) = 1$$

Hence the total increment in z is 1.

(c)   The gradient of PQ is given by substituting into (6.8):

$$G = \frac{-2 + 3}{\sqrt{(4 + 36)}} = \frac{1}{\sqrt{40}} \simeq 0.1581$$

The equation of a straight line in three-dimensional space

In two dimensions, the equation of a straight line takes the form $y = a + bx$, but in three dimensions the analogous equation $z = a + bx + cy$ represents a plane surface in geometric terms. The question naturally arises: 'What form of equation describes a straight line in three-dimensional space?'

Two plane surfaces intersect in a straight line, and any straight line can be represented as the intersection of two appropriate planes. Let the two planes be specified by the equations

$$z - c_1 y - b_1 x = a_1$$

$$z - c_2 y - b_2 x = a_2$$

Subtracting the first equation from the second gives

$$z \quad - c_1 y \quad\quad - b_1 x = a_1$$

$$(c_1 - c_2)y + (b_1 - b_2)x = a_2 - a_1$$

Two equations in three unknowns with a non-zero right-hand side is a degenerate inhomogeneous system (page 24); hence there is no unique solution. Put $x = \lambda$, then the second equation above gives

$$y = \frac{a_2 - a_1}{c_1 - c_2} - \frac{b_1 - b_2}{c_1 - c_2} \lambda$$

and substitution into the first equation yields

$$z = \left\{ a_1 + \frac{c_1(a_2 - a_1)}{c_1 - c_2} \right\} + \lambda \left\{ b_1 - \frac{c_1(b_1 - b_2)}{c_1 - c_2} \right\}$$

The solution may be put into vector form:

$$\begin{bmatrix} x \\ y \\ z \end{bmatrix} = \lambda \begin{bmatrix} 1 \\ \dfrac{b_2 - b_1}{c_1 - c_2} \\ b_1 - \dfrac{c_1(b_1 - b_2)}{c_1 - c_2} \end{bmatrix} + \begin{bmatrix} 0 \\ \dfrac{a_2 - a_1}{c_1 - c_2} \\ a_1 + \dfrac{c_1(a_2 - a_1)}{c_1 - c_2} \end{bmatrix}$$

or

$$\mathbf{x} = \lambda \mathbf{p} + \mathbf{q} \tag{6.9}$$

Hence an equation of the form of (6.9), where each column vector contains three elements, defines a straight line in three-dimensional space.

Example 6.2 Find the equation of the straight line of intersection of the two planes:

$$z = 1 + 3x + 2y$$
$$z = 4 + x + 5y$$

Also find the gradient of this line with respect to the horizontal plane.

First, rearrange the equations into the usual form for solution:

$$z - 3x - 2y = 1$$
$$z - x - 5y = 4$$

Subtract the first equation from the second:

$$z - 3x - 2y = 1$$
$$2x - 3y = 3$$

Put $x = \lambda$, then the second equation yields $y = \frac{2}{3}\lambda - 1$; substitution into the first equation then gives $z = \frac{13}{3}\lambda - 1$. So the vector equation of the straight

191

line of intersection is

$$\begin{bmatrix} x \\ y \\ z \end{bmatrix} = \lambda \begin{bmatrix} 1 \\ \frac{2}{3} \\ \frac{13}{3} \end{bmatrix} + \begin{bmatrix} 0 \\ -1 \\ -1 \end{bmatrix}$$

To find the gradient of the line, find the co-ordinates of any two points on the line, and then substitute in (6.8) with Δz in the numerator:

put $\lambda = 1$; then $x = 1$, $y = -\frac{1}{3}$, $z = \frac{10}{3}$

put $\lambda = 2$; then $x = 2$, $y = \frac{1}{3}$, $z = \frac{23}{3}$

So, $\Delta x = 1$, $\Delta y = \frac{2}{3}$, and $\Delta z = \frac{13}{3}$. Substituting into (6.8) gives

$$G = \frac{4.\dot{3}}{\sqrt{\{(1)^2 + (0.\dot{6})^2\}}} \simeq 3.6$$

Thus the gradient of the straight line with respect to the x–y (horizontal) axial plane is 3.6. The situation is depicted in Figure 6.2.

DIRECTION COSINES

In some applications it is more convenient to describe the orientation of a direction in three-dimensional space in a different way from the above. Referring again to Figure 6.2, we can define the following relationships:

$$\cos Q\hat{P}A = \frac{PA}{PQ}, \qquad \cos Q\hat{P}B = \frac{PB}{PQ}, \qquad \cos Q\hat{P}C = \frac{PC}{PQ}$$

These three quantities, which are called **direction cosines,** define the orientation of line PQ completely, and also have other useful properties.

Firstly, we have

$$PQ^2 = PA^2 + PB^2 + PC^2$$

i.e.

$$PQ^2 = PQ^2 \cos^2 Q\hat{P}A + PQ^2 \cos^2 Q\hat{P}B + PQ^2 \cos^2 Q\hat{P}C$$

and so

$$\cos^2 Q\hat{P}A + \cos^2 Q\hat{P}B + \cos^2 Q\hat{P}C = 1 \qquad (6.10$$

The notation is often abbreviated:

$$l = \cos Q\hat{P}A, \qquad m = \cos Q\hat{P}B, \qquad n = \cos Q\hat{P}C$$

Thus

$$l^2 + m^2 + n^2 = 1$$

Secondly, if two lines have direction cosines (l, m, n) and (l', m', n'), th

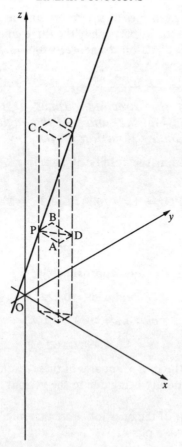

Figure 6.2 A straight line in three-dimensional space. The basis of the concept of direction cosines is also shown.

angle between them θ, is given by

$$\cos \theta = ll' + mm' + nn' \tag{6.11}$$

In particular, the two lines are perpendicular to one another if

$$ll' + mm' + nn' = 0 \tag{6.12}$$

Thirdly, direction cosines enable one to specify the *sense of direction*. In relation to the situation shown in Figure 6.2, we have defined the cosines as positive quantities corresponding to the direction PQ. Corresponding direction cosines for the direction QP would be given as $(-l, -m, -n)$.

Fourthly, two parallel lines have sets of direction cosines which are *numerically* equal to one another, but one set may be reversed in sign.

193

Fifthly, a line at right angles to any co-ordinate axis has a corresponding direction cosine of zero. In particular, the three co-ordinate axes (or lines parallel to them) have direction cosines $(1, 0, 0)$, $(0, 1, 0)$, $(0, 0, 1)$ for the x-, y-, and z-axes, respectively.

Example 6.3   Find the direction cosines of the straight line of Example 6.2 (PQ in Fig. 6.2) and find the angle of this line with a line in the corresponding direction on the x–y plane.

In Figure 6.2 and from the results of Example 6.2, $PA = 1$, $PB = 0.\dot{6}$, $PC = 4.\dot{3}$. Thus

$$PQ^2 = (1)^2 + (0.\dot{6})^2 + (4.\dot{3})^2 = 20.\dot{2}$$

and so

$$PQ \simeq 4.4969$$

The direction cosines are thus, approximately,

$$l = 1/4.4969 = 0.2224$$

$$m = 0.\dot{6}/4.4969 = 0.1483$$

$$n = 4.\dot{3}/4.4969 = 0.9636$$

It will be found that the sum of squares of these results add to 0.999 979 61, the discrepancy from unity being due to the various rounding errors in the above stages.

For the second part of the question, note that line PD is parallel with the specified line in the x–y plane.

Now

$$PD^2 = (1)^2 + (0.6)^2 = 1.\dot{4}$$

and so

$$PD \simeq 1.2019$$

The direction cosines of line PD are approximately

$$l' = 1/1.2019 = 0.8321$$

$$m' = 0.\dot{6}/1.2019 = 0.5547$$

$$n' = 0 \text{ exactly}$$

Hence, if θ is the angle between lines PQ and PD,

$$\cos \theta = (0.2224)(0.8321) + (0.1483)(0.5547) + (0.9636)(0)$$

$$= 0.2673$$

and so $\theta = 74.5°$. Tan $74.5° = 3.6$, and so the result agrees with the mode of working in Example 6.2.

Geometrical representations of functions of two variables

Two methods of graphical presentation of functions of two variables are commonly employed, and these methods are shown for the linear function in Figure 6.3. The plane surface in Figure 6.3a is the same as that already depicted in Figure 6.1 and used in Example 6.1; its equation is $z = 21 - x + \frac{1}{2}y$. A grid of lines, which are parallel to the x–z and y–z axial planes, respectively, has been drawn 'on the surface', and the pattern of this grid emphasises the flatness of the plane surface. In this and all subsequent three-dimensional graphs, the surface is shown only as it occurs in the octant formed by the co-ordinate axes in which all values of the variables are positive. Hence edges of the surface appear when it intersects axial planes.

An alternative way of depicting a function of two variables is to employ two dimensions only, based on the x–y axial plane. We already know that a plane surface intersects the x–y axial plane as a straight line (eqn(6.4)), and for the example in hand the equation of this line is

$$y = -42 + 2x$$

Example 6.1(a)). This line, whose gradient is 2, is drawn as the bold line towards the right-hand side of Figure 6.3b. The label $z = 0$ shows that the line actually lies on the x–y plane because $z = 0$ everywhere on that plane.

Another interpretation of the line $y = -42 + 2x$ is that it shows the relationship between x and y in the overall function $z = 21 - x + \frac{1}{2}y$ when $= 0$. Similarly, we can show the relationship between x and y for any other value of z. In Figure 6.3b we show the situation for $z = 10$ for which we have $10 = 21 - x + \frac{1}{2}y$, i.e. $y = -22 + 2x$, also a straight line with a gradient of 2; and further the situation for $z = 20$ for which we have $0 = 21 - x + \frac{1}{2}y$, i.e. $y = -2 + 2x$, another straight line with a gradient of . These last two lines are thus shown *projected* onto the x–y plane; they can be thought of as contour lines on the surface at different heights (z-values) above the x–y plane.

The above two methods of graphical presentation will be employed for all the functions of two variables to be specifically described in this chapter. With regard to the three-dimensional graphs, it is important to realise that the distance between all lines of the grid drawn on the surface is the same. However, apart from the linear (plane) surface just discussed, this will not *appear* to be the case for the other surfaces to be described because of the way the effects of perspective interact with curvature in different directions.

195

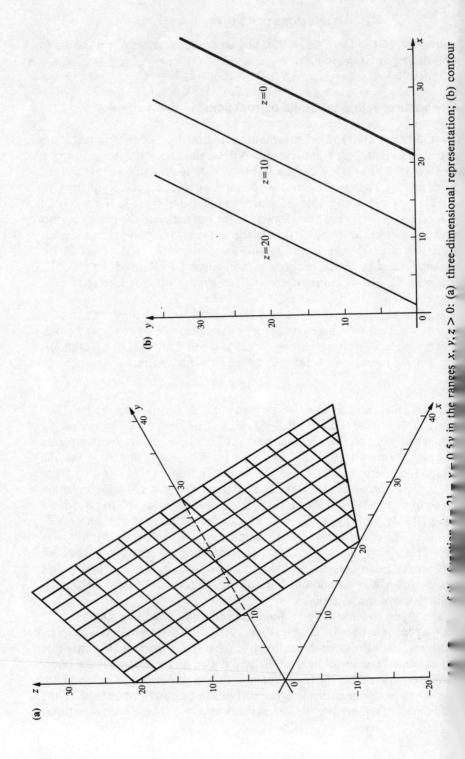

Fig. ... function $z = 21 + x - 0.5 y$ in the ranges $x, y, z > 0$: (a) three-dimensional representation; (b) contour

(a)

(b)

$z=0$

$z=10$

$z=20$

However, this will be a help rather than a hindrance, because the distortions of the regular grid assist the eye to perceive the surface as it really is, even though it is drawn on a two-dimensional sheet of paper. On the two-dimensional representation of the surface, the contour curve at $z = 0$ will be shown in bold, and other examples of contours (at other z-values) will be drawn as ordinary curves.

Conic sections

The idea of representing a function of two variables in two dimensions by means of contour curves projected onto the $x-y$ plane is so important that we need to have some idea of the kinds of curve that can arise before proceeding to a systematic description of non-linear functions of two variables.

In two dimensions, the most general relationship between variables x and y of the second degree must contain the following terms and their coefficients: a constant, and terms in x, x^2, y, y^2, and xy. The relationship is usually written as

$$ax^2 + 2hxy + by^2 + 2gx + 2fy + c = 0 \qquad (6.13)$$

and, depending on the interrelationships among the coefficient values in a specific case, appears geometrically as one of the following:

(a) no real curve at all (see (2) below),
(b) two straight lines,
(c) a circle,
(d) an ellipse,
(e) a parabola,
(f) a hyperbola.

Equation (6.13) is known as the general equation of a conic, because each of the possible geometric representations (cases (b) to (f) above) are sections of a full double cone (Fig. 6.4). The doubling of some of the coefficients merely serves to simplify the notation in some manipulations of the equation and in the results produced.

The following results will be stated without proof.

(1) If (6.13) factorises into two linear factors, it represents two intersecting straight lines (case (b) above). The test for this is

$$D = \begin{vmatrix} a & h & g \\ h & b & f \\ g & f & c \end{vmatrix} = 0$$

197

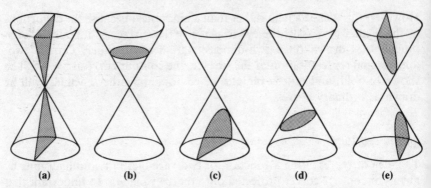

Figure 6.4 Conic sections: (a) two interesting straight lines; (b) circle; (c) parabola; (d) ellipse; (e) hyperbola.

(2) If $D > 0$ and $h^2 - ab < 0$, then no real curve is produced (case (a) above).

To see how this condition may arise consider any particular real value of x, X, say. Now rearrange (6.13) into a quadratic in y:

$$by^2 + 2(hX + f)y + (aX^2 + 2gX + c) = 0$$

If the discriminant

$$4(hX + f)^2 - 4b(aX^2 + 2gX + c) < 0$$

then no real values of y will correspond to X. If a set of a-, c-, f-, g-, and h-values are such that *any* real value of X will result in a negative discriminant, this means there can never be any real values of y and so no line or curve in the real x–y plane can exist.

(3) If $D < 0$, $a = b$ and $h = 0$, we have a circle, centre $(-g/a, -f/b)$.

(4) If $D < 0$ and $h^2 - ab < 0$, equation (6.13) represents an ellipse.

(5) If $h^2 - ab = 0$, we have a parabola.

(6) If $h^2 - ab > 0$, the curve is a hyperbola.

(7) If $h^2 - ab \neq 0$, that is the curve is not a parabola, then the conic section has a centre whose co-ordinates are given by

$$x = (bg - fh)/(h^2 - ab) \text{ and } y = (af - gh)/(h^2 - ab)$$

(8) If the curve is an ellipse it has two axes at right angles to one another coinciding with the longest and shortest diameters. The squares of the half-lengths (radii) of these axes are given by

$$r^2 = \frac{-c}{2(ab - h^2)} [(a + b) \pm \sqrt{\{(a - b)^2 + 4h^2\}}] \qquad (6.14$$

Both values of r^2 are positive quantities, and the radii themselves ar given by the positive square root of each value of r^2 obtained from

(6.14). Doubling the results then gives the lengths of the axes. If the axes of the ellipse are parallel to the co-ordinate axes, then $h = 0$ and (6.14) reduces to $r = (-c/a)$ and $r = (-c/b)$. The shape and orientation of an ellipse are determined by the magnitudes of parameters a, b, c, and h; the position of the ellipse relative to the co-ordinate axes (i.e the position of the centre) is also partly determined by the magnitudes of f and g of (6.12). If $f = g = 0$, the ellipse is centred at the origin.

(9) If the curve is a hyperbola it has two straight lines associated with it which are asymptotes. The rectangular hyperbola, in which the asymptotes are at right angles to each other, is introduced in Causton (1983). The condition for asymptotes to be at right angles to one another, in the notation of (6.13), is $a + b = 0$; however, in a general hyperbola there is no need for the asymptotes to be at right angles.

Quadratic functions

The equation of the second degree polynomial in two dimensions (function of one variable) is

$$y = a + bx + cx^2 \tag{6.15}$$

Extended to three dimensions, as a function of two variables, we have

$$z = a + bx + cx^2 + dy + ey^2 \tag{6.16}$$

but this is not the most general form, which should contain an xy term. However, we shall omit such a term at present for simplicity, but we shall introduce it later on (page 208).

In the case of the parabola produced by relationship (6.15), the curve has a minimum point and is open towards high positive values of y if $c > 0$, and the curve has a maximum point and is open towards high negative values of y if $c < 0$ (see Causton 1983, Fig. 4.1). Clearly the latter case is the more biologically useful. Imagine x to represent an environmental factor (or external factor) and y to represent a certain measurement made on organisms (or the biological response). The curve and equation then represent the way the response changes with levels of the external factor. Now, in several situations the response, y, rises to a maximum and then falls over a range of increasing levels of the external factor, x; for example, plant cell elongation in response to auxin concentration or the abundance of a certain plant species in relation to soil pH in an ecosystem. However, a biological situation in which the response would decrease to a minimum and then increase with changing levels of the external factor must be very rare, if indeed such a situation occurs at all.

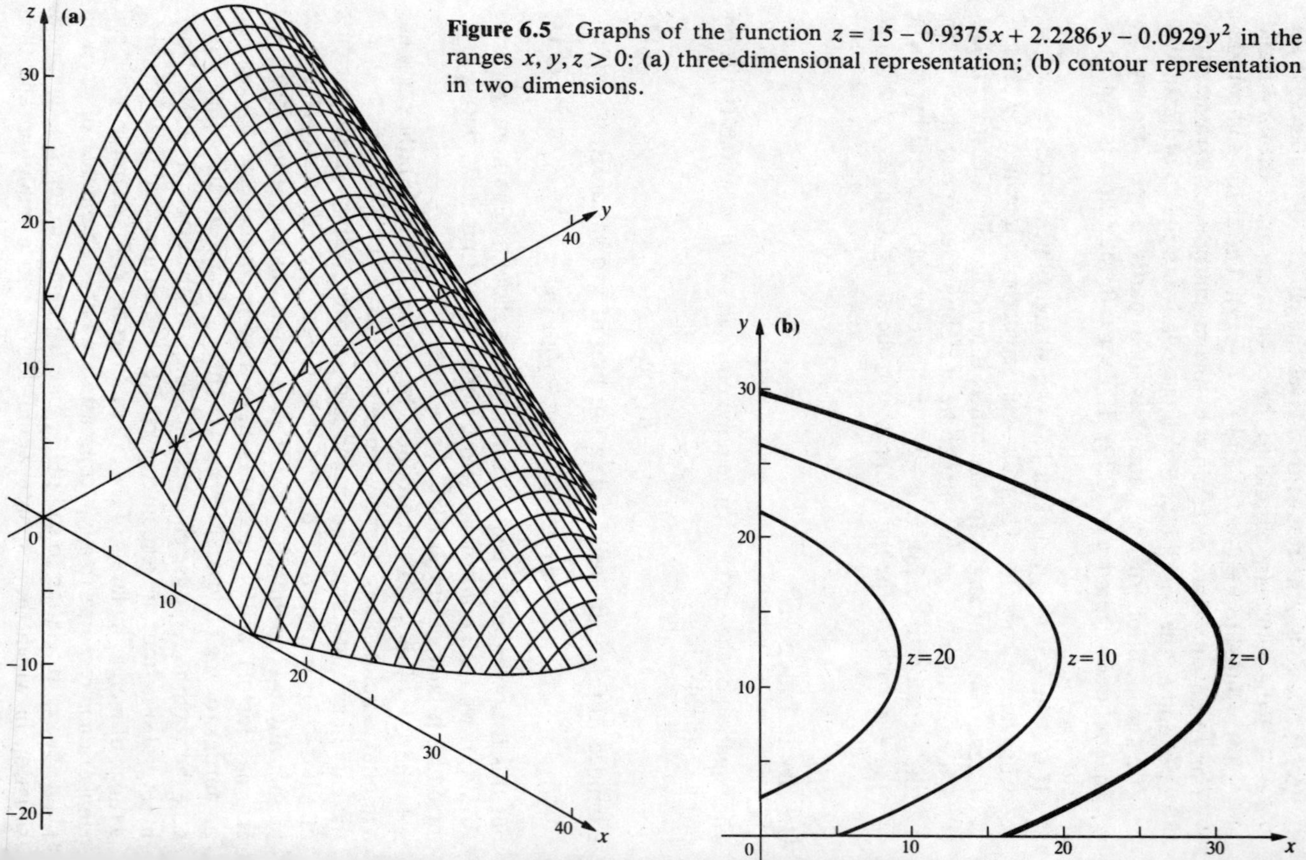

Figure 6.5 Graphs of the function $z = 15 - 0.9375x + 2.2286y - 0.0929y^2$ in the ranges $x, y, z > 0$: (a) three-dimensional representation; (b) contour representation in two dimensions.

In view of the above discussion, we shall illustrate the various forms of (6.16), mostly assuming that coefficients c and e are negative; however, at this stage the main purpose of examining specific examples of quadratic functions of two variables is to introduce mathematical concepts, rather than to review biological applicability.

The function $z = a + bx + dy + ey^2$

In the function

$$z = a + bx + dy + ey^2 \qquad (6.17)$$

the coefficient of x^2 is zero, and so the cx^2 term of (6.16) disappears. Hence z is linearly dependent on x but dependent in a quadratic manner on y. The surface shown in Figure 6.5a, besides depicting the form were e is negative, also has b negative. Because the function is linear in one variable and quadratic in the other, the surface is a tunnel running downhill (or uphill, depending on the way you look at it!) and, as expected, the surface intersects the x–z axial plane in a straight line whose equation is $z = a + bx$ ($b < 0$ here); it intersects the y–z axial plane in a parabola $z = a + dy + ey^2$ ($e < 0$ in this example). For any value of y, z is linearly dependent on x, represented by the straight lines on the surface, and the gradient is always b; conversely, for any value of x, z depends parabolically on y as shown by the curved lines on the surface, and the maximum is always at $y = d/(2e)$.

To investigate the curve of intersection of the surface with the x–y plane, and other contour curves, put $z = 0$ and rearrange (6.17) into the form of (6.13):

$$ey^2 + dy + bx + a = 0 \qquad (6.18)$$

In terms of the coefficients of (6.13), $a = 0$ and $h = 0$; thus $h^2 - ab = 0$ and so relationship (6.18) is a parabola (result (5) on page 198). It can be re-arranged in a more familiar form as

$$x = -\frac{a}{b} - \frac{d}{b}y - \frac{e}{b}y^2 \qquad (6.19)$$

and is shown as the bold curve in Figure 6.5b. Similar curves for other constant values of z are also shown, and in each case the maximum value of x is at $y = -d/(2e)$ – at the same y-value as for maxima of the parabolae in the y–z direction.

Example 6.4   *The equation of the surface shown in Figure 6.5a is*

$$z = 15 - 0.9375x + 2.2286y - 0.0929y^2 \qquad (6.20)$$

What is:

(a) the equation of the line of intersection of the surface with a vertical plane, parallel to the x–z axial plane, at $y = 4$;

(b) the equation of the curve of intersection of the surface with a vertical plane, parallel to the y–z axial plane, at $x = 20$;

(c) the y co-ordinate of the apex of the surface and the maxima of the contour parabolae?

(a) The required equation is obtained by putting $y = 4$ in (6.20):

$$z = 15 - 0.9375x + (2.2286)(4) - (0.0929)(16)$$

i.e.

$$z = 22.4280 - 0.9375x$$

(b) The required equation is obtained by putting $x = 20$ in (6.20):

$$z = 15 - (0.9375)(20) + 2.2286y - 0.0929y^2$$

i.e.

$$z = -3.75 + 2.2286y - 0.0929y^2$$

(c) With respect to (6.17), the point we require is at $y = -d/2e$. Hence

$y = -2.2286/\{2(-0.0929)\} = 12.00$ to two decimal places.

The function $z = a + bx + cx^2 + dy + ey^2$

In the function

$$z = a + bx + cx^2 + dy + ey^2 \tag{6.21}$$

z is parabolically dependent on both x and y, and the precise form of the surface primarily depends on the signs of coefficients c and e.

COEFFICIENTS c AND e BOTH NEGATIVE

When both the coefficients of the squares of the variables are negative, the surface takes the form shown in Figure 6.6a and is called a **paraboloid.** Any vertical section parallel to either the x–z or the y–z axial plane intersects the surface in a parabola, as can be appreciated from the grid lines drawn on the surface in Figure 6.6a. For any value of x, the parabolic curve relating y and z can be shown to have the same position of maximum at $y = -d/(2e)$, although the value of the maximum will depend on the value of x; conversely, for any value of y, the quadratic curve relating x and z has the same position of maximum at $x = -b/(2c)$, and again the value of the maximum depends on the value of y. The apex of the paraboloid is over the point $\{-b/(2c), -d/(2e)\}$ on the x–y plane. By substituting in (6.21),

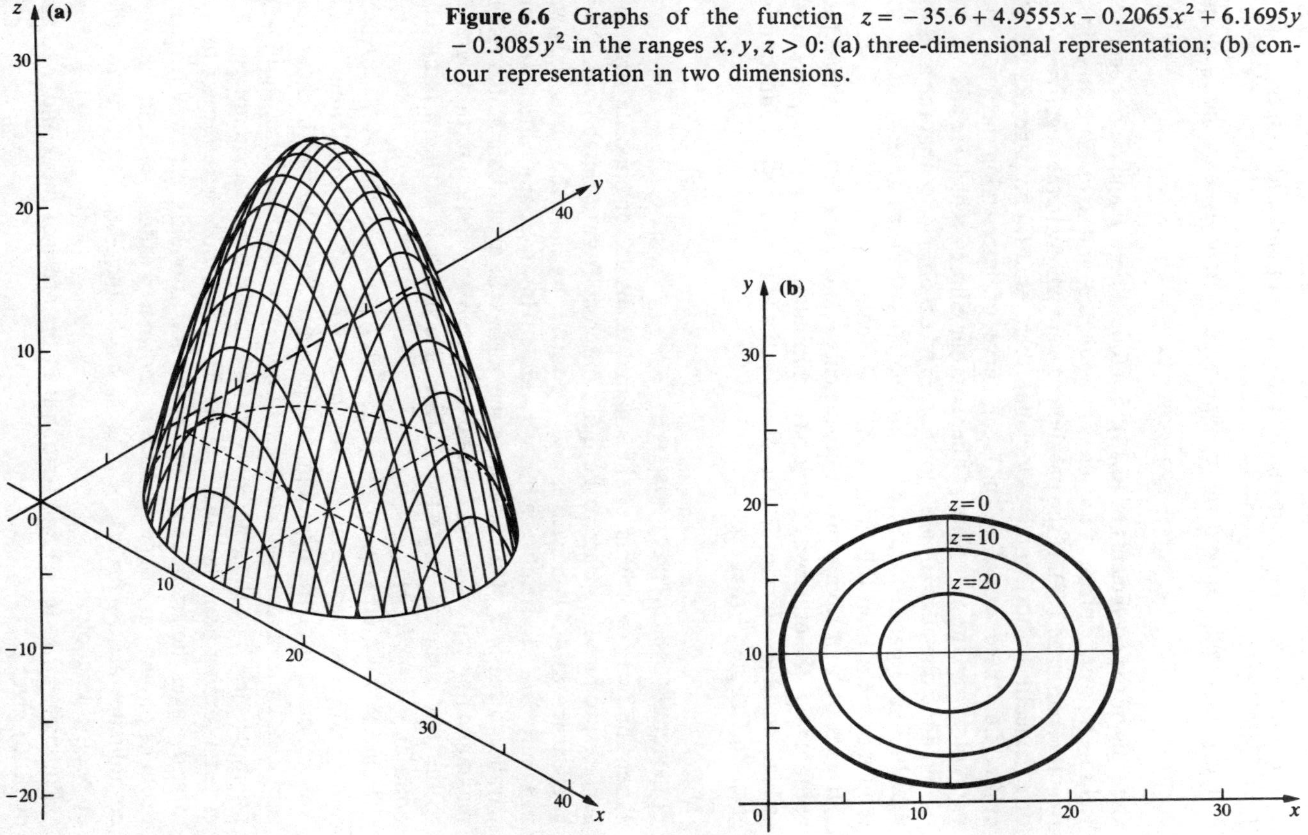

Figure 6.6 Graphs of the function $z = -35.6 + 4.9555y - 0.2065x^2 + 6.1695y - 0.3085y^2$ in the ranges $x, y, z > 0$: (a) three-dimensional representation; (b) contour representation in two dimensions.

we find that the co-ordinates of the apex are $\{-b/(2c), -d/(2e),$ $a-b^2/(4c)-d^2/(4e)\}$, and this is a maximum point with respect to any horizontal direction.

The paraboloid intersects the $x-y$ plane in a curve given by the equation

$$cx^2 + ey^2 + dy + bx + a = 0 \qquad (6.22)$$

In terms of the coefficients of (6.13), $h = 0$, and both a and b are negative, yielding a positive product ab. Thus $h^2 - ab < 0$, and so (6.22) is an ellipse (result (4) on page 198). If the paraboloid were wholly in the negative range of z (beneath the horizontal $x-y$ plane), then result (2) on page 198 would apply, i.e. there would be no real curve of intersection between the paraboloid and the $x-y$ plane. Other contour ellipses, which are all concentric with the one at $z = 0$, are shown in Figure 6.6b. Notice that the axes of the ellipse are parallel with the co-ordinate axes; in the absence of an xy term in the equation this is necessarily the case.

It goes without saying that if coefficients c and e are both positive, the paraboloid is the other way up; it has a minimum point, and is open ended at high positive values of z. An example of the use of such a function is given in Chapter 7 (page 239).

COEFFICIENT c NEGATIVE, e POSITIVE

When the two coefficients of the squared terms are of opposite sign, the surface produced appears to be quite different from a paraboloid. With coefficient c in (6.21) negative, and e positive, we have the surface shown in Figure 6.7a: it resembles a horse's saddle. However, any vertical section parallel to the $x-z$ plane intersects the surface in a parabola which has a maximum point at $x = -b/(2c)$, and conversely any vertical section parallel to the $y-z$ plane intersects the surface in a parabola having a minimum point at $y = -d/(2e)$.

Point A on the $x-y$ plane (Fig. 6.7a) has co-ordinates $\{-b/(2c),$ $-d/(2e)\}$. The surface above this point, B is at the maximum in the x-direction, but at the minimum in the y-direction. Overall, it cannot be said to be either a maximum or a minimum point, only that it is a **stationary point,** because here the gradient is zero in all directions. It cannot even be called a turning point because in two certain directions with respect to the x- and y-axes the gradient remains horizontal as one travels away from B. From its physical analogue, point B is known as a **saddle point,** and its full co-ordinates are $\{-b/(2c), -d/(2e), a-b^2/(4c)-d^2/(4e)\}$ as for the apex of a paraboloid.

Putting $z = 0$ in (6.21) and writing it in the order of the x and y terms of (6.13) gives

$$cx^2 + ey^2 + dy + bx + a = 0 \qquad (6.23)$$

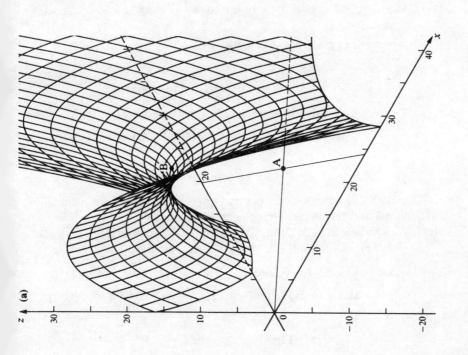

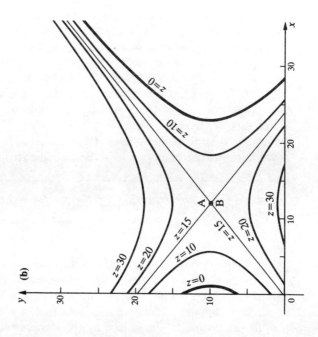

Figure 6.7 Graphs of the function $z = 15.9 + 2.9801x - 0.1242x^2 - 3.7546y + 1877y^2$ in the ranges $x, y, z > 0$: (a) three-dimensional representation; (b) contour representation in two dimensions.

as in (6.22). Equation (6.23) is, of course, that of the curve of intersection of the surface with the x–y plane; in terms of the coefficients of (6.13), $h = 0$ and the product ab is negative because these coefficients of the squared terms are of opposite signs. Thus $h^2 - ab > 0$, and so (6.23) is the equation of a hyperbola (result (6) on page 198). The curve is shown in Figure 6.7b ($z = 0$), together with curves for other values of z; it is evident that when $z > 15$ (for this particular example) the curves are orientated differently from those when $z < 15$. When $z = 15$, the contour curves become two straight lines (result (1) on page 197); these are, in fact, the asymptotes of the family of hyperbolae generated by the surface.

Example 6.5 The equation of the surface shown in Figure 6.6a is

$$z = -35.6 + 4.9555x - 0.2065x^2 + 6.1695y - 0.3085y^2$$

What are:
(a) *the equation of the curve of intersection of the surface with a vertical plane, parallel to the x–z axial plane, at y = 6;*
(b) *the equation of the curve of intersection of the surface with a vertical plane, parallel to the y–z axial plane, at x = 12;*
(c) *the co-ordinates and nature of the stationary point?*

(a)   The required equation is obtained by putting $y = 6$ in the above equation:

$$z = -35.6 + 4.9555x - 0.2065x^2 + (6.1695)(6) - (0.3085)(36)$$

i.e.

$$z = -9.6890 + 4.9555x - 0.2065x^2$$

(b)   The required equation is obtained by putting $x = 12$ in the equation:

$$z = -35.6 + (4.9555)(12) - (0.2065)(144) + 6.1695y - 0.3085y^2$$

i.e.

$$z = -5.87 + 6.1695y - 0.3085y^2$$

Both the equations in (a) and (b) are parabolae with maximum points and their open ends facing 'downwards'.

(c)   With respect to (6.21), the stationary point has co-ordinates $\{ -b/(2c),$ $-d/(2e), a - b^2/(4c) - d^2/(4e)\}$. Hence we have

$$x = -4.9555/\{2(-0.2065)\}, \quad y = -6.1695/\{2(-0.3085)\}$$

$$z = -35.6 - (4.9555)^2/\{4(-0.2065)\} - (6.195)^2/\{4(-0.3085)\}$$

and so

$$x = 12, \quad y = 10, \quad \text{and} \quad z = 25$$

Because the coefficients of both the x^2 and the y^2 terms are negative, the stationary point is a maximum.

Example 6.6   *The equation of the surface shown in Figure 6.7a is*

$$z = 15.9 + 2.9801x - 0.1242x^2 - 3.7546y + 0.1877y^2$$

What are:
(a)   the equation of the curve of intersection of the surface with a vertical plane, parallel to the x–z axial plane, at y = 18;
(b)   the equation of the curve of intersection of the surface with a vertical plane, parallel to the y–z axial plane, at x = 1;
(c)   the co-ordinates and nature of the stationary point?

(a)   The required equation is obtained by putting $y = 18$ in the above equation:

$$z = 15.9 + 2.9801x - 0.1242x^2 - (3.7546)(18) + (0.1877)(324)$$

i.e. $\qquad\qquad z = 9.1320 + 2.9801x - 0.1242x^2$

which is a parabola with a maximum point and open end pointing 'downwards'.

(b)   The required equation is obtained by putting $x = 1$ in the above equation:

i.e.

$$z = 15.9 + (2.9801)(1) - (0.1242)(1) - 3.7546y + 0.1877y^2$$
$$z = 18.76 - 3.7546y + 0.1877y^2$$

which is a parabola with a minimum point and open end facing 'upwards'.

(c)   With respect to (6.21), the stationary point has co-ordinates $\{ -b/(2c),$ $-d/(2e),\ a - b^2/(4c) - d^2/(4e)\}$. Hence we have

$$x = -2.9801/\{2(-0.1242)\}, \quad y = -(-3.7546)/\{2(0.1877)\}$$
$$z = 15.9 - (2.9801)^2/\{4(-0.1242)\} - (-3.7546)^2/\{4(0.1877)\}$$

and so

$$x = 12, \quad y = 10, \quad z = 15$$

Because the coefficient of x^2 is negative and that of y^2 is positive, the stationary point is a saddle point.

207

Functions with interactions between the variables

In the functions considered so far in this chapter, the two variables are independent; that is, in the function $z = f(x, y)$, any *change* in z occasioned solely by a particular *change* in x is the same for any value of y, and conversely any change in z occasioned only by a particular change in y is the same regardless of the value of x selected. For example, in the linear function $z = 21 - x + \frac{1}{2}y$ (Example 6.1), z decreases one unit for every unit increase in x and a fixed y regardless of the actual value of y; and, conversely, z increases one half unit for every unit increase in y and fixed x regardless of the latter's value. We say that the two variables are **independent,** or that there is no **interaction** between the variables.

In many biological situations, however, variables may interact to produce a response. In terms of the above discussion, this means that the response, z, to a change in x only at a fixed value of y depends on what value y actually is; or, put another way, the change in z occasioned by a unit increase in x at one fixed y-value is different from the change in z occasioned by a unit increase in x at another fixed value of y. Mathematically, a convenient way of accounting for interaction is to introduce an xy term in the function, and we shall examine two of the functions previously discussed but now with an interaction term present.

The 'linear' function with interaction

The 'linear' function with interaction is obtained by adding an xy term to the right-hand side of (6.1):

$$z = a + bx + cy + fxy \qquad (6.24)$$

The surface is shown in Figure 6.8a, and it is evident from the graph that the surface is curved and no longer plane. Equation (6.24) also shows that we are no longer dealing with a linear function since such a function cannot contain the product of two variables. There is no generally accepted name for the function described by (6.24), but 'the "linear" function with interaction' is acceptable if the word 'linear' is placed in quotation marks; this name is useful since it is a precise description of the origin of the function.

When $y = 0$, the surface intersects the x–z axial plane in a straight line whose equation is $z = a + bx$; conversely, when $x = 0$, the surface intersects the y–z axial plane in a straight line whose equation is $z = a + cy$. From Figure 6.8a it is evident that at any fixed value of y there is a linear relationship between x and z, but the gradient of the straight line differs depending on the fixed value of y.

In the example shown in Figure 6.8a, whose equation is

$$z = 28 - x - 1.2y + 0.1xy$$

Figure 6.8 Graphs of the function $z = 28 - x - 1.2y + 0.1xy$ in the range $x, y, z > 0$: (a) three-dimensional representation; (b) contour representation in two dimensions.

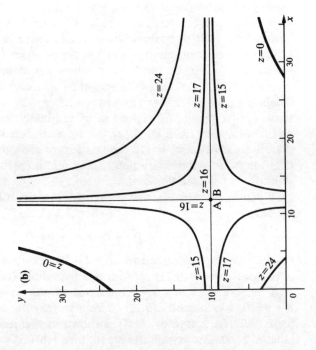

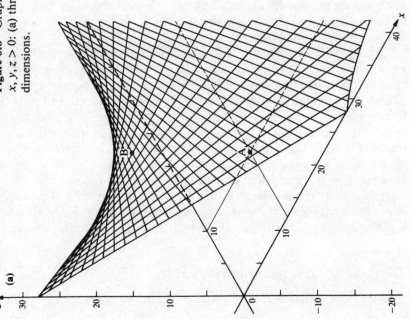

the negative gradient of the x–z relationship decreases in numerical value as y increases from zero; the gradient becomes zero at $y = 10$, and then becomes increasingly positive as y increases above 10. Similarly the gradient of the linear relationship between y and z increases from an initial negative value when $x = 0$, becoming zero when $x = 12$ and positive for higher x-values. When $y = 10$, $z = 28 - x - 12 + x$, i.e. $z = 16$, showing that z is independent of x; similarly, when $x = 12$, $z = 28 - 12 - 1.2y + 1.2y = 16$, showing that now z is independent of y. Finally, above the point $(12, 10)$ on the x–y plane (A in Fig 6.8a), $z = 16$; and, since we have already shown that when $x = 12$ and $y = 10$ the gradient of the surface is zero, the point $(12, 10, 16)$ is a stationary point (B in Fig. 6.8a). Moreover, Figure 6.8a shows it to be a saddle point.

When $z = 0$ in (6.24), and the terms are written in the order of (6.13), we have

$$fxy + cy + bx + a = 0 \qquad (6.25)$$

and in terms of the coefficients of (6.13), $a = b = 0$ and $h = f$. Thus $h^2 - ab$ is positive and so (6.25) represents a hyperbola (result (6) on page 198). Furthermore since $a = b = 0$ (in terms of the coefficients of eqn(6.13)), $a + b = 0$, which shows (6.25) to be a rectangular hyerbola (result (9) on page 199). The curve of (6.25) is shown in Figure 6.8b as the bold lines labelled $z = 0$. The asymptotes are the lines labelled $z = 16$ (compare Figs 6.8 a & b and the present discussion with that in the previous paragraph). Some contour curves at other z-values are also shown in Figure 6.8b, projected down onto the x–y plane.

To summarise all the above in biological terms, although in the function $z = a + bx + cy + fxy$ any change in z (a response) occasioned solely by a change in x (level of an external factor) is linear, the amount of change in z for a given change in x depends on the prevailing fixed value of y. In the linear case, $z = a + bx + cy$, the amount of change in z for a given change in x is the same for any fixed value of y. Any experimentally derived response relationship is only valid over the ranges of the external variables employed in the experiment; obviously linear response relationships can only apply over strictly limited ranges of values of the external factors whether interaction is present or not. Thus a linear function describing a relationship between a biological response and one or more external factors is biologically highly artificial but nevertheless mathematically convenient.

The quadratic function with interaction

The quadratic function with interaction has the equation

$$z = a + bx + cx^2 + dy + ey^2 + fxy \qquad (6.26)$$

and we shall consider just one form of this which is biologically the most useful, namely where coefficients c and e are both negative (cf. page 202). Writing (6.26) with $z = 0$ and the order of the terms on the right-hand side in line with those on the left-hand side of (6.13), we have

$$cx^2 + fxy + ey^2 + dy + ex + a = 0 \qquad (6.27)$$

In terms of the coefficients of (6.13), $h^2 - ab$ (i.e. in terms of our present coefficients $f^2/4 - ce$) could be any value, thus making (6.27) represent an ellipse, a parabola, or a hyperbola, depending on the relative values of c, e and f (notation of (6.27)). If f is small in relation to c and e, then (6.27) is an ellipse and (6.26) a paraboloid, and this is the situation to be considered in this section.

As we saw in the previous section, introducing an xy term into a linear function may cause quite a pronounced change in the geometrical properties of the surface. However, the introduction of an interaction term into a paraboloid alters the geometry relatively little. A graphical presentation of (6.26) where c and e are less than zero (negative) and $f = 0$ (no xy term) has already been given in Figure 6.6. In terms of the contour ellipses (Fig. 6.6b), the only change occasioned by the presence of an xy term is to tilt the axes of the ellipses in the $x-y$ plane so that they are no longer parallel with the co-ordinate axes. Viewing the situation in another way, we can say that the intoduction of an xy term in the equation of a paraboloid causes a rotation of the contour ellipses about their common centre to a new position. Extending this idea to the paraboloid itself (Fig. 6.6a), we see that the introduction of an xy term merely affects the orientation of the surface in relation to the xy plane; there is no change of shape at all.

The interaction term, however, does have one important consequence. In the absence of an xy term, the maximum of any parabolic section in a vertical plane, parallel to the $x-z$ axial plane, is always at $x = -b/(2c)$, regardless of the value of y at which the vertical plane is situated (page 202); a similar statement can be made for the parabola in the $y-z$ direction at any fixed value of x. For (6.26) it can be shown that the maximum of a parabolic section in the $x-z$ direction is at

$$x = \frac{-b - fy}{2c}$$

the position of the maximum with respect to the x-direction depending on the value of y. Similarly, the maximum of a parabolic section in the $y-z$ direction is at

$$y = \frac{-d - fx}{2e}$$

the position of the maximum with respect to the y-direction depending on the value of x. Hence, where a response to two external factors takes the form of a paraboloid, and there is interaction between the variables, the principal effect of the interaction is that one variable influences the position of maximum response to the other variables.

Partial differentiation

In the equation of a straight line in two-dimensional space, one of the constants, or parameters, of the equation specifies the constant gradient. In three-dimensional space we have seen, in relation to a plane surface, that there is no unique gradient; the gradient of a surface (involving change in z-value) can only be specified with respect to a particular direction in the x–y plane.

For a plane surface, $z = a + bx + cy$, the gradient in the x-direction is b and in the y-direction is c; as in the analogous straight line, the constant gradients in the x- and y-directions are specified directly in the equation of the plane. Further, equation (6.8) gives the gradient of the plane in any desired direction.

Curved surfaces do not usually have constant gradients in any direction, but there are exceptions (see Figs 6.5a & 6.8a). We now require an extension to the rules of differentiation to deal with functions of more than one variable. This extension is known as **partial differentiation,** and the clue to its operation is provided by the fact that we can only find the gradient of a surface when a particular direction has been specified.

First derivatives

In a function of two variables, $z = f(x, y)$, we can differentiate z with respect either to x or y in the one operation while holding the other variable constant. To show that we are dealing with a function of more than one variable, and that we are differentiating only with respect to one of them we write $\partial z/\partial x$ or $\partial z/\partial y$ instead of $\mathrm{d}z/\mathrm{d}x$ or $\mathrm{d}z/\mathrm{d}y$.

Example 6.7 Differentiate the following functions with respect to each variable in turn:

(a) $z = a + bx + cy$ (b) $z = a + bx + cx^2 + dy + ey^2 + fxy$

(a) We have directly $\partial z/\partial x = b$, since the terms a and cy are constants from the point of view of x. Similarly, $\partial z/\partial y = c$. These results confirm what we already know about the plane surface, that the gradient in the x-direction is b and that in the y-direction is c.

(b) Again, we have directly,

$$\frac{\partial z}{\partial x} = b + 2cx + fy$$

and

$$\frac{\partial z}{\partial y} = d + 2ey + fx$$

Second derivatives

Just as in the case of ordinary differentiation of a function of a single variable, so may second partial derivatives be obtained. The notation is $\partial^2 z/\partial x^2$ and $\partial^2 z/\partial y^2$, respectively, for a function $z = f(x, y)$, but an additional higher derivative also exists: $\partial^2 z/(\partial x\,\partial y)$.

To clarify the geometrical meanings of these higher derivatives, we can write

$$\frac{\partial^2 z}{\partial x^2} = \frac{\partial(\partial z/\partial x)}{\partial x} \tag{6.28}$$

The right-hand side of (6.28) shows that the second derivative $\partial^2 z/\partial x^2$ is the rate of change of the rate of change of z with respect to x, at constant y. What about the mixed higher derivative? We write

$$\frac{\partial^2 z}{\partial x\partial y} = \frac{\partial(\partial z/\partial x)}{\partial y} \tag{6.29}$$

which implies that (6.29) is the rate of the change of $\partial z/\partial x$ with respect to y, or the rate of change of gradient in the x-direction with respect to y. We can also write

$$\frac{\partial^2 z}{\partial x\,\partial y} = \frac{\partial(\partial z/\partial y)}{\partial x} \tag{6.30}$$

which implies that the mixed derivative can also be interpreted as rate of change of gradient in the y-direction with respect to x. For most functions, (6.29) and (6.30) are equal to one another.

The meaning of the mixed derivative can be clarified still further by considering the case of the 'linear' function with interaction.

Example 6.8   Obtain the first and second partial derivatives of

$$z = a + bx + cy + fxy$$

and explain their meanings.

213

We have

$$\frac{\partial z}{\partial x} = b + fy \qquad \frac{\partial z}{\partial y} = c + fx$$

$$\frac{\partial^2 z}{\partial x^2} = 0 \qquad \frac{\partial^2 z}{\partial y^2} = 0$$

$$\frac{\partial^2 z}{\partial x \, \partial y} = f$$

The last result follows regardless of whether we differentiate $\partial z/\partial x$ with respect to y or $\partial z/\partial y$ with respect to x. The 'pure' second derivatives, $\partial^2 z/\partial x^2$ and $\partial^2 z/\partial y^2$, being zero, express the fact that the rate of change of the rate of change of gradient in the x- or y- direction is zero. The mixed second derivative shows that the rate of change of gradient in the x-direction with respect to rate of change of y is not zero, as can be seen in Figure 6.8a; similarly, the rate of change of gradient in the y-direction with respect to the rate of change x is non-zero. In the absence of an xy term, the mixed derivative is zero, showing that the rate of change of gradient in one direction is zero regardless of any change in the other direction.

Example 6.9   Obtain the first and second partial derivative of

(a) $5x^2y^2 - 7x^3y^5$
(b) $(x^2 - y^2)/(x^2 + y^2)$

(a)   Put z equal to the function; then, treating the ys as constants,

$$\frac{\partial z}{\partial x} = 10xy^3 - 21x^2y^5$$

Treating the xs as constants, we have

$$\frac{\partial z}{\partial y} = 15x^2y^2 - 35x^3y^4$$

Next, differentiate $\partial z/\partial x$ with respect to x:

$$\frac{\partial^2 z}{\partial x^2} = 10y^3 - 42xy^5$$

then $\partial z/\partial y$ with respect to y:

$$\frac{\partial^2 z}{\partial y^2} = 30x^2y - 140x^3y^3$$

and, finally, differentiate either $\partial z/\partial x$ with respect to y or $\partial z/\partial y$ with

214

respect to x:

$$\frac{\partial^2 z}{\partial x \, \partial y} = 30xy^2 - 105x^2y^4$$

(b) Again, put z equal to the function and differentiate with respect to x using the quotient rule:

$$\frac{\partial z}{\partial x} = \frac{2x(x^2 + y^2) - 2x(x^2 - y^2)}{(x^2 + y^2)^2}$$

remembering that y-terms are constants in this case; on simplifying we have

$$\frac{\partial z}{\partial x} = \frac{4xy^2}{(x^2 + y^2)^2}$$

Next, differentiate the function with respect to y:

$$\frac{\partial z}{\partial x} = \frac{-2y(x^2 + y^2) - 2y(x^2 - y^2)}{(x^2 + y^2)^2}$$

i.e.

$$\frac{\partial z}{\partial y} = \frac{-4x^2y}{(x^2 + y^2)^2}$$

To obtain the second partial derivatives, we first differentiate the denominator of the above expressions by the function of a function rule; so

$$\frac{\partial\{(x^2 + y^2)^2\}}{\partial x} = 4x(x^2 + y^2)$$

$$\frac{\partial\{(x^2 + y^2)^2\}}{\partial y} = 4y(x^2 + y^2)$$

Now differentiate $\partial z/\partial x$ with respect to x, using the quotient rule:

$$\frac{\partial^2 z}{\partial x^2} = \frac{4y^2(x^2 + y^2)^2 - 16x^2y^2(x^2 + y^2)}{(x^2 + y^2)^4}$$

and simplifies to

$$\frac{\partial^2 z}{\partial x^2} = \frac{4y^2(y^2 - 3x^2)}{(x^2 + y^2)^3}$$

Next, differentiate $\partial z/\partial y$ with respect to y:

$$\frac{\partial^2 z}{\partial y^2} = \frac{-4x^2(x^2 + y^2)^2 + 16x^2y^2(x^2 + y^2)}{(x^2 + y^2)^4}$$

215

which simplifies to

$$\frac{\partial^2 z}{\partial y^2} = \frac{4x^2(3y^2 - x^2)}{(x^2 + y^2)^3}$$

Finally, differentiate $\partial z/\partial x$ with respect to y:

$$\frac{\partial^2 z}{\partial x \, \partial y} = \frac{8xy(x^2 + y^2)^2 - 16xy^3(x^2 + y^2)}{(x^2 + y^2)^4}$$

and $\partial z/\partial y$ with respect to x:

$$\frac{\partial^2 z}{\partial x \, \partial y} = -\left\{ \frac{8xy(x^2 + y^2)^2 - 16x^3 y(x^2 + y^2)}{(x^2 + y^2)^4} \right\}$$

both of which simplify to

$$\frac{\partial^2 z}{\partial x \, \partial y} = \frac{8xy(x^2 - y^2)}{(x^2 + y^2)^3}$$

The gradient of a surface

The gradient of a surface in the x-direction is given by $\partial z/\partial x$, and the gradient in the y-direction is $\partial z/\partial y$. To find the gradient in any other direction, we need to proceed in a similar manner to the derivation for the plane surface on page 189. Figure 6.1 can again be used, but now we imagine points P, Q, and R to lie on a curved surface, and the intervals Δx, Δy, and Δz to be small and so written as δx, δy, and δz. For the plane surface, (6.7) is then written as

$$\delta z = b \, \delta x + c \, \delta y$$

and, because $\partial z/\partial x = b$ and $\partial z/\partial y = c$ for the plane surface, (6.7) becomes

$$\delta z = \frac{\partial z}{\partial x} \, \delta x + \frac{\partial z}{\partial y} \, \delta y \tag{6.31}$$

We shall now see that where the intervals are sufficiently small equation (6.31) holds approximately for any surface (it is exact for the plane surface). The method is similar to that in the 'Small increments' section of Chapter 5, but now we are applying the Taylor series in two dimensions instead of one.

Imagine two points on a surface close together; P is any point with co-ordinates (x, y, z) and Q is distant from P δx in the x-direction, δy in the y-direction, and δz in the z-direction, this last being dependent on the values of x and δx, y and δy, and the equation of the surface $z = f(x, y)$. In fact, we have

$$\delta z = (z + \delta z) - z$$

216

or

$$\delta z = f(x + \delta x, y + \delta y) - f(x, y) \qquad (6.32)$$

We now require to express this increment in terms of the derivatives of z at the point $P(x, y, z)$.

First, in the function $f(x + \delta x, y + \delta y)$ we expand $x + \delta x$ as a Taylor series in x, keeping $y + \delta y$ constant:

$$f(x + \delta x, y + \delta y) = f(x, y + \delta y) + \frac{\partial \{f(x, y + \delta y)\}}{\partial x} \delta x$$

$$+ \tfrac{1}{2} \frac{\partial^2 \{f(x, y + \delta y)\}}{\partial x^2} (\delta x)^2 + \cdots \qquad (6.33)$$

Now expand $f(x, y + \delta y)$ as a Taylor series in y, keeping x constant:

$$f(x, y + \delta y) = f(x, y) + \frac{\partial \{f(x, y)\}}{\partial y} \delta y + \tfrac{1}{2} \frac{\partial^2 \{f(x, y)\}}{\partial y^2} (\delta y)^2 + \cdots$$

but $f(x, y) = z$ by definition, and so the latter expansion becomes

$$f(x, y + \delta y) = z + \frac{\partial z}{\partial y} \delta y + \tfrac{1}{2} \frac{\partial^2 z}{\partial y^2} (\delta y)^2 + \cdots \qquad (6.34)$$

Substituting (6.34) for $f(x, y + \delta y)$ in (6.33), and substituting for $f(x + \delta x, y + \delta y)$ in (6.32), we have

$$\delta z \simeq z + \frac{\partial z}{\partial y} \delta y + \tfrac{1}{2} \frac{\partial^2 z}{\partial y^2} (\delta y)^2 + \frac{\partial [z + (\partial z/\partial y)\, \delta y + \tfrac{1}{2}\, (\partial^2 z/\partial y^2)(\delta y)^2]}{\partial x} \delta x$$

$$+ \tfrac{1}{2} \frac{\partial^2 [z + (\partial z/\partial y)\, \delta y + \tfrac{1}{2}\, (\partial^2 z/\partial y^2)(\delta y)^2]}{\partial x^2} (\delta x)^2 - z$$

where we use $\simeq$ because we have eliminated all terms above second order. Expanding, and observing that the z-terms cancel,

$$\delta z \simeq \frac{\partial z}{\partial y} \delta y + \tfrac{1}{2} \frac{\partial^2 z}{\partial y^2} (\delta y)^2 + \frac{\partial z}{\partial x} \delta x + \frac{\partial^2 z}{\partial x\, \partial y} \delta x\, \delta y$$

$$+ \tfrac{1}{2} \frac{\partial^3 z}{\partial x\, \partial y^2} \delta x (\delta y)^2 + \tfrac{1}{2} \frac{\partial^2 z}{\partial x^2} (\delta x)^2$$

$$+ \tfrac{1}{2} \frac{\partial^3 z}{\partial x^2\, \partial y} (\delta x)^2\, \delta y + \tfrac{1}{4} \frac{\partial^4 z}{\partial x^2\, \partial y^2} (\delta x)^2 (\delta y)^2$$

Finally, eliminating all terms above second order again and rearranging we arrive at

$$\delta z \simeq \frac{\partial z}{\partial x} \delta x + \frac{\partial z}{\partial y} \delta y + \tfrac{1}{2} \left\{ \frac{\partial^2 z}{\partial x^2} (\delta x)^2 + 2 \frac{\partial^2 z}{\partial x\, \partial y} \delta x\, \delta y + \frac{\partial^2 z}{\partial y^2} (\delta y)^2 \right\} \qquad (6.35)$$

This result may be written as

$$\delta z = \frac{\partial z}{\partial x} \delta x + \frac{\partial z}{\partial y} \delta y + O\{(\delta x)^2\} \tag{6.36}$$

which may be regarded as exact because all terms of second and higher order are included in $O\{(\delta x)^2\}$. If δx and δy are sufficiently small, then all terms of $O\{(\delta x)^2\}$ are negligible and the approximation

$$\delta z \simeq \frac{\partial z}{\partial x} \delta x + \frac{\partial z}{\partial y} \delta y \tag{6.37}$$

is good. Often this approximation is written in the form

$$dz = \frac{\partial z}{\partial x} dx + \frac{\partial z}{\partial y} dy$$

but this is unrigorous and confusing and should be avoided (cf. Ch. 5, page 162). In (6.37), δx, δy, and δz imply small but not vanishingly small intervals, as dx, dy, and dz would imply.

Stationary points

Equation (6.35) shows how a surface changes (in terms of a small interval in the z-direction, δz) in the immediate neighbourhood of a point $P(x, y, z)$. The small intervals, $\delta x, \delta y, \delta z$, must be known, and the partial derivatives are calculated at the given point P. If a point on the surface is such that

$$\partial z / \partial x = \partial z / \partial y = 0$$

at that point, it is a stationary point (see page 204); the expression of second partial derivatives in the braces of (6.35) then determines the shape of the function in the neighbourhood of the stationary point. The following scheme is relevant:

$$\frac{\partial^2 z}{\partial x^2} > 0, \quad \frac{\partial^2 z}{\partial x^2} \frac{\partial^2 z}{\partial y^2} - \left(\frac{\partial^2 z}{\partial x \, \partial y}\right)^2 > 0 \text{ for a minimum,}$$

$$\frac{\partial^2 z}{\partial x^2} < 0, \quad \frac{\partial^2 z}{\partial x^2} \frac{\partial^2 z}{\partial y^2} - \left(\frac{\partial^2 z}{\partial x \, \partial y}\right)^2 > 0 \text{ for a maximum,}$$

$$\frac{\partial^2 z}{\partial x^2} \frac{\partial^2 z}{\partial y^2} - \left(\frac{\partial^2 z}{\partial x \, \partial y}\right)^2 < 0 \text{ for a saddle point.}$$

If the point in question is a saddle point, then there are two lines in the $x-y$ direction at which z remains constant. At points on these lines, known as **level lines,** the second-order terms in the braces of (6.35) also vanish; the gradients of the two level lines (dy/dx) are given as the roots of the quadratic equation in dy/dx:

$$\left(\frac{dy}{dx}\right)^2 \frac{\partial^2 z}{\partial y^2} + 2\left(\frac{dy}{dx}\right)\frac{\partial^2 z}{\partial x\,\partial y} + \frac{\partial^2 z}{\partial x^2} = 0 \qquad (6.38)$$

If $\partial^2 z/\partial x^2 + \partial^2 z/\partial y^2 = 0$ the level lines are perpendicular. Further, as a point moves around the saddle point, in the sense of a projection down onto the x–y plane, z is successively greater than, less than, greater than, less than the z-value of the saddle point as the level lines are crossed.

Example 6.10   Investigate the stationary points of

(a) $z = a + bx + cx^2 + dy + ey^2$ with $c, e < 0$

(b) $z = a + bx + cx^2 + dy + ey^2$ with $c < 0, e > 0$

(c) $z = a + bx + dy + fxy$

(a)   Differentiating, we have

$$\frac{\partial z}{\partial x} = b + 2cx \qquad\qquad \frac{\partial z}{\partial y} = d + 2ey$$

$$\frac{\partial^2 z}{\partial x^2} = 2c \qquad\qquad \frac{\partial^2 z}{\partial y^2} = 2e$$

$$\frac{\partial^2 z}{\partial x\,\partial y} = 0$$

Because $c < 0$, $\dfrac{\partial^2 z}{\partial x^2} < 0$; $e < 0$, $\dfrac{\partial^2 z}{\partial y^2} < 0$

Hence the product $(\partial^2 z/\partial x^2)(\partial^2 z/\partial y^2) > 0$, and because $\partial^2 z/\partial x\,\partial y = 0$, we have the conditions for a maximum point. This checks with what we already know about this function (page 204 and Fig. 6.6a). To find the co-ordinates in the x–y plane of this maximum point, equate $\partial z/\partial x$ and $\partial z/\partial y$ to zero; thus

$$b + 2cx = 0$$

and

$$d + 2ey = 0$$

giving

$$x = -b/2c \quad\text{and}\quad y = -d/2e.$$

(b) The same derivatives apply as in (a) above. Because $\partial^2 z/\partial x^2 < 0$ and $\partial^2 z/\partial y^2 > 0$, their product is negative, and so with $\partial^2 z/\partial x\,\partial y = 0$ we have a saddle point. The saddle point has the same x- and y-co-ordinates as the maximum point in (a) above. Applying equation (6.38) to obtain the gradients of the level lines, we get

$$2e\left(\frac{\mathrm{d}y}{\mathrm{d}x}\right)^2 + 2c = 0$$

so

$$\frac{\mathrm{d}y}{\mathrm{d}x} = \pm \sqrt{(-c/e)}$$

and since c is negative and e is positive the square root is real. Finally, in general

$$\frac{\partial^2 z}{\partial x^2} + \frac{\partial^2 z}{\partial y^2} = 2(c + e) \neq 0$$

although if c and e are numerically equal this expression is zero. So in general the level lines are not at right angles to one another, but if c and e are numerically equal to one another, then the level lines are at right angles to each other.

(c) Here we have

$$\frac{\partial z}{\partial x} = b + fy \qquad\qquad \frac{\partial z}{\partial y} = d + fx$$

$$\frac{\partial^2 z}{\partial x^2} = 0 \qquad\qquad \frac{\partial^2 z}{\partial y^2} = 0$$

$$\frac{\partial^2 z}{\partial x\,\partial y} = f$$

The stationary point is where $b + fy = d + fx = 0$, and so $y = -b/f$ and $x = -d/f$. Since both $\partial^2 z/\partial x^2 = \partial^2 z/\partial y^2 = 0$, and $(\partial^2 z/\partial x\,\partial y)^2 > 0$ whatever the value of f, we have a saddle point. Applying equation (6.38), we have

$$2f\frac{\mathrm{d}y}{\mathrm{d}x} = 0$$

giving a gradient of zero for one of the level lines with respect to the x–y plane. Further, since $\partial^2 z/\partial x^2 + \partial^2 z/\partial y^2 = 0$, the two level lines are at right angles to one another, and because of the previous result they are parallel to the x- and y-coordinate axes.

Suggestions for further reading

In the context of this chapter, further reading would be required to obtain a deeper insight into the theory of partial differentiation. Most standard textbooks on calculus, or such general mathematical texts as Stephenson (1961) or Heading (1970) (referenced in earlier chapters), will provide the necessary material.

Exercises

6.1 Obtain all the first and second partial derivatives of the function $z = e^{-kx} \cos ay$.
6.2 Find the stationary points on the surface $z = x^3 + xy + y^2$ and determine their nature.

7

Fitting functions to data

Having surveyed the properties of several functions of one or more variables, we now turn to the problems involved in fitting functions to data. The main use of a mathematical function to a biologist is to describe succinctly a set of acquired experimental or survey data; and a summary of the uses to which a fitted mathematical function, describing a set of data, are put is given in Causton (1983, pp. 40–2).

Biological, and most natural environmental, data are always subject to random variation of greater or lesser magnitude; hence, even if it could be shown theoretically that a particular type of function underlies a specific set of data, the actual observations would deviate from the curve to a greater or lesser extent. Almost invariably, however, we do not know that a particular kind of function is theoretically correct for the data in hand, although we may sometimes have a hypothetical function in mind as a result of the construction of a **mechanistic model** (page 246). More usually we are using a mathematical function as an **empirical model**: a set of data seems to accord with the shape of a curve of a particular function, so we fit that function; we are essentially summarising the data by means of a convenient mathematical relationship. This immediately raises the question of how large are the deviations of the data from the fitted curve, which in turn leads on to the question of how good is the fit. Statistical methods are required to answer such questions and, further, if a good fit is obtained there may be one or more hypotheses to test concerning the fitted function – again involving statistical methods.

Broadly speaking, therefore, the topics of **curve fitting**, or **function estimation**, is statistical in nature; and full accounts may be found in various statistical textbooks, particularly in Sprent (1969) and the ever-popular Draper and Smith (1981). However, the actual fitting process itself, based on the **principle of least squares**, is largely devoid of statistical ideas. The application of the principle of least squares is not always straightforward, however, and there is now an extensive literature on function estimation. The material in this chapter is intended as an introduction to the subject.

The straight line

To illustrate the general principles of the method of least squares we start with the simplest case – fitting a straight line. First, we need to define the nature of the data. Very often, observations requiring the fitting of a function are univariate (a biological measurement) and represented by the y-axis of a graph. We are usually interested in the influence of an external factor, x, on the biological measurement, y. The term 'external' means a factor external to the biological system being studied under experimental conditions, such as time, temperature, light intensity, particle size of a substratum, etc; the values of the levels of an external factor should be known relatively precisely because they are determined by the experimenter, whereas the biological measurement is subject to random variation. Hence all the random variation of the system occurs in the vertical direction of a usual $x-y$ graph, as shown in Figure 7.1. This further implies that we are concerned with the magnitudes of the deviations of the data from the fitted line solely in the vertical direction, that is, the length of the vertical lines in Figure 7.1.

It can be appreciated that any fitted straight line will have deviations from it to the data points but that different fitted straight lines to the same set of data will have deviations of different magnitudes. Intuitively, it is

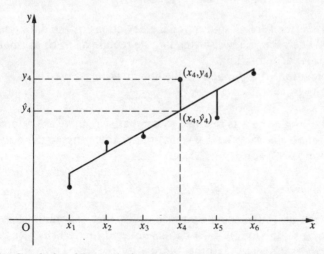

Figure 7.1 Graph showing a typical set of data to which a straight line is fitted. The x-axis represents a factor external to the biological system under study (typically an environmental factor), and y is a biological measurement made (biological response). The line is the best fitting straight line, and the length of the vertical line from each datum point to the fitted line represents the deviation of that point from the line.

223

apparent that the best fitting straight line is one in which the deviations as a whole are minimal; so to establish this criterion we must first establish an expression quantifying a set of deviations, then we can minimise that expression.

Referring again to Figure 7.1, we symbolise the co-ordinates of each datum point as (x_i, y_i), $i = 1, \ldots, n$ where n is the total number of data items (six in this case). For any particular x-value, x_i, there are two relevant y-values: the datum value of y, y_i, and the y-value given by the fitted straight line, $\hat{y}_i$. This is shown specifically for $i = 4$ in Figure 7.1. Evidently for any one x-value the deviation of the datum point from the fitted line is $(y_i - \hat{y}_i)$. Points above the line will thus have positive deviations, and points below the line have negative deviations.

At first sight, it may be thought appropriate to measure total deviations as their sum, i.e.

$$\sum_{i=1}^{n} (y_i - \hat{y}_i)$$

but owing to the presence of a mixture of positive and negative deviations this sum could be zero although individual deviations may be very large. To overcome this problem, we actually define

$$S^2 = \sum_{i=1}^{n} (y_i - \hat{y}_i)^2 \tag{7.1}$$

where S^2 is 'the sum of squares of the deviations of the data points about the fitted line'. For the best fitting line we require S^2 to be as small as possible, that is, at minimum.

The equation of a *fitted* straight line is written as

$$\hat{y} = a + bx \tag{7.2}$$

where the $\hat{}$ over the y is retained to emphasise that we are referring to the straight line value of y at a given x rather than any actual datum value. Substituting (7.2) for y_i in (7.1) gives

$$S^2 = \sum_{i=1}^{n} (y_i - a - bx_i)^2 \tag{7.3}$$

and for the best fitting line expression (7.3) must be minimised.

In fitting a function, it must be remembered that a fit is being sought to *a particular set of data*; in other words, for the set of data in hand, the x_i and y_i in (7.3) are constants. Further, finding the best fitting line may be regarded as asking the question, what value of a and of b make (7.3) a minimum? *So as regards fitting a function to a particular set of data we have the reverse situation to normal, that in (7.3) the x_i and y_i are the constants and a and b are the variables.*

224

Equation (7.3) is thus a function of two variables, a and b; so to minimise the expression we must partially differentiate successively with respect to a and b and then equate the resulting pair of equations to zero. Differentiation in the presence of a $\sum$ sign presents no additional difficulties; we may ignore the sign when differentiating, but re-insert it into the answer. Further, since we are always fitting to all n data points, the subscripts i of the x and y, together with the limits of summation, may be removed and we can write (7.3) more simply as

$$S^2 = \sum (y - a - bx)^2 \qquad (7.4)$$

To differentiate (7.4), ignoring the summation sign, we employ the function of a function (or chain) rule giving $-2(y - a - bx)$ and $-2(y - a - bx)x$ on differentiating with respect to a and b, respectively. Inserting the $\sum$ signs, we have

$$\frac{\partial S^2}{\partial a} = -2\sum (y - a - bx) \qquad (7.5)$$

and

$$\frac{\partial S^2}{\partial b} = -2\sum (y - a - bx)x \qquad (7.6)$$

Now set the two partial derivatives to zero, and in so doing we may eliminate the constant factors of -2, giving

$$\sum (y - a - bx) = 0$$
$$\sum (y - a - bx)x = 0$$

Next the brackets must be removed and, remembering that it is the x and y terms that are summed and that in this context the a and b are constants, we have

$$\sum y - an - b\sum x = 0$$
$$\sum xy - a\sum x - b\sum x^2 = 0$$

or

$$an + b\sum x = \sum y \qquad (7.7)$$
$$a\sum x + b\sum x^2 = \sum xy \qquad (7.8)$$

The term an comes from what at first sight is $\sum a$ or $a\sum$. Bearing in mind that there are n items (of unity) in this summation, the result is an.

Equations (7.7) and (7.8) are a pair of non-degenerate inhomogeneous linear equations in a and b, which may be easily solved for particular data sets. However, they may also be solved symbolically. Dividing throughout

(7.7) by n and rearranging gives

$$a = \bar{y} - b\bar{x} \qquad (7.9)$$

where $\bar{x} = (\sum x)/n$ and $\bar{y} = (\sum y)/n$, the means of the data values of x and y, respectively. Substituting into (7.8):

$$(\bar{y} - b\bar{x})\sum x + b\sum x^2 = \sum xy$$

i.e.

$$\bar{y}\sum x - b\bar{x}\sum x + b\sum x^2 = \sum xy$$

i.e.

$$\frac{\sum x \sum y}{n} + b\left(\sum x^2 - \frac{\sum^2 x}{n}\right) = \sum xy$$

and so

$$b = \frac{\sum xy - (\sum x \sum y)/n}{\sum x^2 - (\sum^2 x)/n} \qquad (7.10)$$

which is equivalent to

$$b = \frac{\sum (x - \bar{x})(y - \bar{y})}{(x - \bar{x})^2} \qquad (7.11)$$

Note that $\sum^2 x = (\sum x)^2$.

Equations (7.9) and (7.10) or (7.11) are, of course, the well known results for calculating the best fitting straight line to a set of data by least squares.

The minimised deviations sum of squares, S^2, is an important quantity, and it can be calculated by evaluating $\hat{y}_i$ at each x_i and then applying equation (7.1). However, in the case of a fitted straight line, a more convenient formula for the deviation sum of squares can be derived. Substituting for a in (7.4) we have

$$S^2 = \sum \{y - (\bar{y} - b\bar{x}) - bx\}^2$$
$$= \sum \{(y - \bar{y}) - b(x - \bar{x})\}^2$$
$$= \sum (y - \bar{y})^2 - 2b\sum (x - \bar{x})(y - \bar{y}) + b^2 \sum (x - \bar{x})^2$$

On substituting for b in the third term above, using (7.11), we have

$$S^2 = \sum (y - \bar{y})^2 - 2b\sum (x - \bar{x})(y - \bar{y}) + \frac{b\sum (x - \bar{x})(y - \bar{y})\sum (x - \bar{x})^2}{\sum (x - \bar{x})^2}$$

and so

$$S^2 = \sum (y - \bar{y})^2 - b\sum (x - \bar{x})(y - \bar{y}) \qquad (7.12)$$

which, again, is a well known result.

226

Equations (7.7) and (7.8) can also be written (and solved) in matrix form:

$$\begin{bmatrix} n & \sum x \\ \sum x & \sum x^2 \end{bmatrix} \begin{bmatrix} a \\ b \end{bmatrix} = \begin{bmatrix} \sum y \\ \sum xy \end{bmatrix}$$

The nature of the pair of equations can now be readily seen, particularly the symmetry of the matrix of coefficients on the left-hand side.

Some general considerations

The straight line example of the previous section includes all the steps involved in fitting *any* function to data by least squares. These steps are:

(1) Formulate an expression for the sum of squares of the deviations of the data points from a fitted function (eqns (7.3) or (7.4) in the above example).
(2) Partially differentiate the expression established in step 1 with respect to each parameter in turn of the function being fitted.
(3) Equate the resulting partial derivatives to zero.
(4) Solve the set of equations obtained in step 3.

Whatever function is being fitted to data the first three steps can be worked through in the same way as for the straight line, although the differentiation procedures in step 2 can get complicated. Nevertheless, one can always proceed as far as step 3 and obtain a set of equations whose number is the same as the number of function parameters being estimated. It is the last step where differences between different function types become apparent, according to whether the equations are analytically solvable (as are linear equations) or solvable only by numerical methods.

For rather devious historical reasons, the term **regression** is commonly used for function estimation and, according to whether the least squares equations obtained at the end of step 3 are linear or not, two categories of regression are recognised: **linear regression** and **non-linear regression**. In general, linear equations are directly solvable whereas non-linear equations need to be solved numerically.

It must not be thought that linear regression is confined to the straight-line situation. Any function, *linear in its parameters*, will give rise to a linear regression situation. Thus the fitting of the second-degree polynomial

$$\hat{y} = a + bx + cx^2 \tag{7.13}$$

(and indeed a polynomial of any degree) gives rise to a linear regression situation because the parameters a, b, and c appear linearly in the function (7.13). This means that formulae are obtainable, analagous to (7.9) and

227

(7.10) for the straight line, for estimates of the three parameters. However, even for the second-degree polynomial the formulae are much more complicated than they are for the straight line, and the order of complexity increases very rapidly with increase in the degree of polynomial. Hence such formulae are now rarely quoted, and polynomial curve fitting is almost invariably accomplished by a computer routine.

Linear regression: a trigonometric function

To illustrate that a *curvilinear function* can be fitted in a *linear regression* situation, provided the function is linear in its parameters, we exemplify the trigonometric function

$$\hat{y} = a \cos x + b \sin x \qquad (7.14)$$

which has been used as an empirical model to describe body rhythms, where x is time and y is a measured rate of some physiological attribute whose rhythm is of interest.

The function to be minimised is

$$S^2 = \sum (y - a \cos x - b \sin x)^2 \qquad (7.15)$$

Differentiating successively with respect to a and b gives

$$\frac{\partial S^2}{\partial a} = -2\sum (y - a \cos x - b \sin x)\cos x$$

$$\frac{\partial S^2}{\partial b} = -2\sum (y - a \cos x - b \sin x)\sin x$$

Equating to zero, eliminating the -2 factor, multiplying out the brackets, and rearranging gives

$$a\sum \cos^2 x \qquad + b\sum \sin x \cos x = \sum y \cos x$$
$$a\sum \sin x \cos x + b\sum \sin^2 x \qquad = \sum y \sin x$$

In matrix form, we have

$$\begin{bmatrix} \sum \cos^2 x & \sum \sin x \cos x \\ \sum \sin x \cos x & \sum \sin^2 x \end{bmatrix} \begin{bmatrix} a \\ b \end{bmatrix} = \begin{bmatrix} \sum y \cos x \\ \sum y \sin x \end{bmatrix}$$

The inverse of the matrix of coefficients on the left-hand side is given by

$$\mathbf{M}^{-1} = \frac{1}{(\sum \cos^2 x)(\sum \sin^2 x) - \sum^2 \sin x \cos x}$$

$$\begin{bmatrix} \sum \sin^2 x & -\sum \sin x \cos x \\ -\sum \sin x \cos x & \sum \cos^2 x \end{bmatrix}$$

and so

$$\begin{bmatrix} a \\ b \end{bmatrix} = \mathbf{M}^{-1} \begin{bmatrix} \sum y \cos x \\ \sum y \sin x \end{bmatrix}$$

yielding

$$a = \frac{(\sum \sin^2 x)(\sum y \cos x) - (\sum \sin x \cos x)(\sum y \sin x)}{(\sum \cos^2 x)(\sum \sin^2 x) - \sum^2 \sin x \cos x} \qquad (7.16)$$

$$b = \frac{(\sum \cos^2 x)(\sum y \sin x) - (\sum \sin x \cos x)(\sum y \cos x)}{(\sum \cos^2 x)(\sum \sin^2 x) - \sum^2 \sin x \cos x} \qquad (7.17)$$

To fit function (7.14) to $x-y$ data, therefore, first requires the calculation of the following sums:

$$\sum \cos^2 x, \sum \sin^2 x, \sum \sin x \cos x, \sum y \sin x, \sum y \cos x$$

Do not forget that x must be in radians.

In itself equation (7.14) is not a very practical function since it oscillates about a mean of $y = 0$, whereas a measured attribute is always positive. It is easy to adjust (7.14) so that the y-values are always positive: one merely needs to add a constant, c, of sufficient magnitude to give

$$\hat{y} = a \cos x + b \sin x + c$$

The function now has three parameters and, although no new principles are involved, the mathematics of the derivations and the resulting formulae for the estimates of a, b, and c are much more complicated than for the two-parameter case (Exercise 7.2). However, if the mean value of y is known or can be assumed, then y-values at each x can be evaluated as deviations from the mean, and (7.14) applies. The following example assumes this situation, and so the data to which the function is to be fitted contain both positive and negative values of y.

Example 7.1   Find the best fitting function of the form

$$\hat{y} = a \cos x + b \sin x$$

to the following data:

x	1	2	3	4	5	6	7
y	−0.75	−3.75	−2.00	0.50	3.90	2.50	0.10

The calculations can be readily performed on a simple scientific calculator which contains the trigonometric functions with radians as the argument.

We have

$$\sum \cos^2 x = \cos^2 1 + \ldots + \cos^2 7 = (0.5403)^2 + \ldots + (0.7539)^2$$

$$= 0.2919 + \ldots + 0.5684 = 3.4432$$

$$\sum \sin^2 x = \sin^2 1 + \ldots + \sin^2 7 = (0.8415)^2 + \ldots + (0.6570)^2$$

$$= 0.7081 + \ldots + 0.4316 = 3.5568$$

$$\sum \sin x \cos x = (\sin 1)(\cos 1) + \ldots + (\sin 7)(\cos 7)$$

$$= 0.4547 + \ldots + 0.4953 = 0.3863$$

$$\sum y \sin x = -0.75 \sin 1 + \ldots + 0.10 \sin 7$$

$$= -0.6311 + \ldots + 0.0657 = -9.0743$$

$$\sum y \cos x = -0.75 \cos 1 + \ldots + 0.1 \cos 7$$

$$= -0.4052 + \ldots + 0.0754 = 6.3906$$

The denominator of both (7.16) and (7.17) is given by

$$(3.4432)(3.5568) - (0.3863)^2 = 12.0976$$

Hence from (7.16),

$$a = \frac{(3.5568)(6.3906) - (0.3863)(-9.0743)}{12.0976} = 2.1687$$

and from (7.17),

$$b = \frac{(3.4432)(-9.0743) - (0.3863)(6.3906)}{12.0976} = -2.7868$$

Thus the curve of best fit to the seven data points is

$$\hat{y} = 2.1687 \cos x - 2.7868 \sin x$$

For each value of x, a $\hat{y}$-value can now be calculated from the above equation. From these, the deviations sum of squares can be evaluated, as shown below:

x	1	2	3	4	5	6	7	
y	-0.75	-3.75	-2.00	0.50	3.90	2.50	0.10	
$\hat{y}$	-1.17	-3.44	-2.54	0.69	3.29	2.86	-0.20	$\sum(y-\hat{y})^2$
$y-\hat{y}$	0.42	-0.31	0.54	-0.19	0.61	-0.36	0.30	1.1919

Thus the deviations sum of squares about the best fitting curve is 1.1919.

Non-linear regression: the rectangular hyperbola

Introductory appraisal

As an example of a non-linear regression situation we will use the rectangular hyperbola of the form

$$y = \frac{ax}{b + x} \qquad (7.18)$$

This function has many biological uses, both as a mechanistic model – the Michaelis–Menten function of enzyme kinetics – and as an empirical model; for example, the photosynthetic rate–light intensity response curve of a leaf. These two uses of function (7.18) have been described in Causton (1983, Chap. 4).

Several curvilinear functions that are commonly used in biology may be transformed into a straight line. Manipulation of (7.18) will yield

$$\frac{1}{y} = \frac{1}{a} + \frac{b}{a}\left(\frac{1}{x}\right) \qquad (7.19)$$

Thus if the *reciprocals* of the x- and y-co-ordinates of points on a rectangular hyperbola (7.18) are plotted, the straight line (7.19) will result having gradient b/a and intercept on the $(1/y)$-axis of $1/a$. A similar, reciprocal, transformation of both x- and y-values of a set of data will have the same effect, enabling one to obtain numerical estimates of b/a and $1/a$ by the straight-line fitting method already described. Two very elementary calculations will then provide estimates of a and b.

One's immediate reaction could well be that this method of fitting (7.18) is fine; indeed, before the widespread availability of electronic computers it was often the only practical way of fitting a rectangular hyperbola, and other functions that could be transformed into a linear form. However, the method does not give the rectangular hyperbola (7.18) of best fit. True, one obtains the best fitting linear form (7.19); but when fitting to data which are subject to deviations from the underlying curve, the line (7.19) of best fit to the reciprocals of the data does not back-transform to the rectangular hyperbola (7.18) of best fit to the original data. What we can say is that although functions (7.18) and (7.19) are *mathematically* equivalent they are not *statistically* equivalent in the context of fitting to deviant data. If the data points did lie exactly on the curve of (7.18), then the parameters a and b would be validly calculated through straight-line regression on the reciprocals of the x- and y-values of the data, but this is never the situation in practice. We shall return to this topic later (page 237).

FITTING FUNCTIONS TO DATA

Least squares estimation of the rectangular hyperbola

The function to be fitted is

$$\hat{y} = \frac{ax}{b+x}$$

and so the sum of squares to be minimised is given by

$$S^2 = \sum \left(y - \frac{ax}{b+x} \right)^2$$

or, on putting over a common denominator,

$$S^2 = \sum \left(\frac{by + xy - ax}{b+x} \right)^2 \qquad (7.20)$$

Partially differentiating with respect to a presents no more of a problem than in the case of the straight line: use the function of a function rule and ignore the summation sign. We get

$$\frac{\partial S^2}{\partial a} = -2 \sum \frac{(by + xy - ax)x}{(b+x)^2} \qquad (7.21)$$

Differentiating with respect to b requires the application of the quotient rule also, and we end up with

$$\frac{\partial S^2}{\partial b} = 2a \sum \frac{(by + xy - ax)x}{(b+x)^3} \qquad (7.22)$$

It is quite evident that when (7.21) and (7.22) are equated to zero they are not directly solvable for a and b; a numerical approach is required, and it is here that the paths of linear and non-linear regression diverge.

SOLVING THE LEAST SQUARES EQUATIONS

We shall use the Newton–Raphson technique, already presented as a method for numerically solving an equation in one unknown (page 181). The formula for one iteration is

$$x_1 = x_0 - f(x_0)/f'(x_0)$$

where x_0 is an initial (guessed) value of the unknown variable, x_1 is the adjusted (hopefully improved) value of the unknown variable, $f(x_0)$ is the functional form of the equation numerically evaluated for $x = x_0$, and $f'(x_0)$ is the derivative of the functional form also evaluated for $x = x_0$. The method can be extended to include m equations in m unknowns based on the following analagous formulation:

$$\mathbf{x}_1 = \mathbf{x}_0 - \mathbf{M}_0^{-1} \mathbf{v}_0 \qquad (7.23)$$

232

In equation (7.23), x_0 is a column vector, of length m, of initial values of the unknowns, and x_1 is the corresponding column vector of adjusted values. The column vector v_0, also of length m, contains the functional form of each equation, evaluated at the initial value of each unknown; and M_0 is a square matrix, of order m, containing the partial derivatives of the functional form of each equation with respect to each unknown, evaluated at the initial value of each unknown. For our rectangular hyperbola, we have

$$
\begin{bmatrix} a_1 \\ \\ b_1 \end{bmatrix} = \begin{bmatrix} a_0 \\ \\ b_0 \end{bmatrix} - \begin{bmatrix} \dfrac{\partial^2 S^2}{\partial a^2_0} & \dfrac{\partial^2 S^2}{\partial a_0 \partial b_0} \\ \\ \dfrac{\partial^2 S^2}{\partial a_0 \partial b_0} & \dfrac{\partial^2 S^2}{\partial b^2_0} \end{bmatrix}^{-1} \begin{bmatrix} \dfrac{\partial S^2}{\partial a_0} \\ \\ \dfrac{\partial S^2}{\partial b_0} \end{bmatrix} \tag{7.24}
$$

The equations we are solving, (7.21) and (7.22) when equated to zero, are already (first) derivatives of the original expression for the deviations sum of squares, so the matrix M_0 contains second derivatives of the original expression. To obtain these, first differentiate (7.21) with respect to a, which gives directly

$$
\frac{\partial^2 S^2}{\partial a^2} = 2\sum \frac{x^2}{(b+x)^2} \tag{7.25}
$$

Next, using the quotient rule, differentiate (7.21) with respect to b, and then (7.22) with respect to a: the result will be the same, namely

$$
\frac{\partial^2 S^2}{\partial a \partial b} = 2\sum \frac{(by + xy - 2ax)x}{(b+x)^3} \tag{7.26}
$$

Finally, differentiate (7.22) with respect to b, again using the quotient rule:

$$
\frac{\partial^2 S^2}{\partial b^2} = -2a\sum \frac{(2by + 2xy - 3ax)x}{(b+x)^4} \tag{7.27}
$$

We now have all the quantities required for evaluating the left-hand side of (7.24). In each of the first and second partial derivatives the data values of x and y are used, together with the initial values, a_0 and b_0, for a and b. Hence, numerical values of all the elements in the matrix and vectors on the right-hand side of (7.24) are obtained and so the adjusted values, a_1 and b_1, may be calculated. These can be used as initial values in a further iteration, and so on in the usual way. The large amount of calculation arises in the numerical evaluation of the various derivatives; the inversion of the square matrix, the matrix multiplication, and the vector subtraction are quite trivial in comparison when only two parameters require estimation.

233

STARTING VALUES

Hopefully, a sequence of iterations from a pair of given (starting) values of a and b will converge to the curve of best fit, as measured by the deviations sum of squares decreasing with successive iterations to a steady value – the minimum. For many reasons it is highly desirable, if not essential, for starting values to be as close as possible to the converged values (see page 242). In the case of the rectangular hyperbola, the parameter estimates given by the linear form (7.19) provide excellent starting values; use of this approach has the advantage that the whole process of estimating the rectangular hyperbola is automatic, involving the following steps.

(1) Obtain the reciprocals of all the x and y data values.
(2) Fit a straight line to these values of $1/x$ and $1/y$ (cf. eqn (7.19)).
(3) From the numerical estimates of $1/a$ and b/a in Step 2, evaluate a and b: these are the starting values.
(4) Using the starting values in equations (7.21), (7.22), (7.25), (7.26), and (7.27), evaluate the elements of the matrices on the right-hand side of (7.24) and hence calculate adjusted values of a and b, together with the deviations sum of squares.
(5) Repeat Step 4 as often as necessary, using the results of the previous iteration as initial values, until there is no further change in the values of a and b within the number of digits being used.

Example 7.2 Find the best fitting rectangular hyperbola of the form

$$\hat{y} = \frac{ax}{b + x}$$

to the following data:

x	0.2	0.4	0.7	2.0	5.0	10.0
y	1.00	1.92	2.19	3.34	3.49	3.95

The reciprocals, or inverses, of the data are:

$1/x$	5.00	2.50	1.4286	0.50	0.20	0.10
$1/y$	1.00	0.5208	0.4566	0.2994	0.2865	0.2532

The relevant calculations for fitting a straight line to the inverse data are:

$$\sum(1/x) = 9.7286 \qquad \sum(1/x)^2 = 33.5909 \qquad \overline{1/x} = 1.6214$$

$$\sum(1/x - \overline{1/x})^2 = \sum(1/x^2 - \sum^2(1/x)/n = 17.8166$$

$$\sum(1/y) = 2.8165 \qquad \sum(1/y)^2 = 1.7156 \qquad \overline{1/y} = 0.4694$$

$$\sum(1/y - \overline{1/y})^2 = \sum(1/y)^2 - \sum{}^2(1/y)/n = 0.3935$$

$$\sum(1/x)(1/y) = 7.1866$$

$$\sum(1/x - \overline{1/x})(1/y - \overline{1/y}) = \sum(1/x)(1/y) - \sum(1/x)\sum(1/y)/n = 2.6198$$

$$\frac{b}{a} = \frac{\sum(1/x - \overline{1/x})(1/y - \overline{1/y})}{\sum(1/x - \overline{1/x})^2} = \frac{2.6198}{17.8166} = 0.1471$$

and

$$1/a = \overline{1/y} - (b/a)(\overline{1/x})$$
$$= 0.4694 - (0.1471)(1.6214) = 0.2309$$

So

$$a = 1/0.2309 = 4.3309$$

and

$$b = (0.1471)(4.3309) = 0.6371$$

The deviations sum of squares cannot be calculated by means of (7.12) since this implies

$$\sum(1/y - \widehat{1/y})^2$$

where $(\widehat{1/y})$-values are given by (7.19). Instead, values of $\hat{y}$ must be calculated from the equation

$$\hat{y} = \frac{4.3309x}{0.6371 + x}$$

at the data values of x, the deviations $(y - \hat{y})$ evaluated, and finally the deviations are squared and summed over the six data points. So we have

x	0.2	0.4	0.7	2.0	5.0	10.0	$\sum(y - \hat{y})^2$
y	1.00	1.92	2.19	3.34	3.49	3.95	
$\hat{y}$	1.0347	1.6704	2.2673	3.2846	3.8414	4.0715	
$y - \hat{y}$	−0.0347	0.2496	−0.0773	0.0554	−0.3514	−0.1215	0.2108

and so the deviations sum of squares is 0.2108.

Using $a = 4.3309$ and $b = 0.6371$ as starting values, the following values of the first and second partial derivatives (eqns (7.21), (7.22), (7.25), (7.26), (7.27)) are calculated. Even if a computer is not being used for the whole fitting program, something more than an ordinary calculator is required here; the summations involved are not straightforward and so, apart from the time-consuming nature of the computations involved, there are abundant opportunities for errors to accrue. At least a programmable calculator is necessary, or better, a microcomputer.

$$\frac{\partial S^2}{\partial a} = 0.6728 \qquad \frac{\partial S^2}{\partial b} = 0.0220$$

$$\frac{\partial^2 S^2}{\partial a^2} = 5.4513 \qquad \frac{\partial^2 S^2}{\partial a \partial b} = -7.4213 \qquad \frac{\partial^2 S^2}{\partial b^2} = 17.4490$$

Thus in the notation of (7.23) and (7.24), we have

$$\mathbf{v}_0 = \begin{bmatrix} 0.6728 \\ 0.0220 \end{bmatrix} \quad \text{and} \quad \mathbf{M}_0 = \begin{bmatrix} 5.4513 & \\ -7.4213 & 17.4490 \end{bmatrix}$$

Inverting $\mathbf{M}_0$ gives

$$\mathbf{M}_0^{-1} = \begin{bmatrix} 0.4358 & \\ 0.1853 & 0.1361 \end{bmatrix}$$

and so, from (7.24) we get

$$\begin{bmatrix} a_1 \\ b_1 \end{bmatrix} = \begin{bmatrix} 4.3309 \\ 0.6371 \end{bmatrix} - \begin{bmatrix} 0.4358 & 0.1853 \\ 0.1853 & 0.1361 \end{bmatrix} \begin{bmatrix} 0.6728 \\ 0.0220 \end{bmatrix}$$

$$= \begin{bmatrix} 4.3309 \\ 0.6371 \end{bmatrix} - \begin{bmatrix} 0.2973 \\ 0.1277 \end{bmatrix} = \begin{bmatrix} 4.0336 \\ 0.5094 \end{bmatrix}$$

Calculating the deviations sum of squares from the function

$$\hat{y} = \frac{4.0336x}{0.5094 + x}$$

by the same method as before, we obtain $S^2 = 0.1184$, a reduction of nearly a half.

Repeating the same calculations, using a_1 and b_1 as initial values gives $a_2 = 4.0600$, $b_2 = 0.5286$, and $S^2 = 0.1160$. A further iteration yields $a_3 = 4.0610$, $b_3 = 0.5296$, and $S^2 = 0.1160$. The deviations sum of squares to four decimal places has not changed, and a further iteration gives $a_4 = a_3$ and $b_4 = b_3$ to four decimal places. Hence the required best fitting rectangular hyperbola to the given set of data is

$$\hat{y} = \frac{4.061x}{0.5296 + x}$$

with deviations sum of squares of 0.1160.

The linear form: a final rebuttal!

Having worked through an example of fitting a rectangular hyperbola to data, we are now in a position to appraise further the undesirability of fitting the linear form to the inversely transformed data.

Figure 7.2a shows the data employed in Example 7.2, together with the fitted curve. You will observe that the deviations of all the data points from the curve, in the vertical direction, are *approximately* the same. Figure 7.2b shows the reciprocals of the same data, together with a straight line which is the linear form of the curve in Figure 7.1a. It is clear that the magnitudes of the deviations are not approximately equal but increase from small values where $1/x$ and $1/y$ are small (x and y large) to larger values where $1/x$ and $1/y$ are large (x and y small).

Now the correct non-linear least squares fitting procedure for the rectangular hyperbola, described above, effectively gives each datum point the same weighting in the fitting process. A linear regression straight line fitting to (7.19) would also give equal weighting to the transformed data points. However, as Figure 7.2b shows, the data are inherently less variable at one

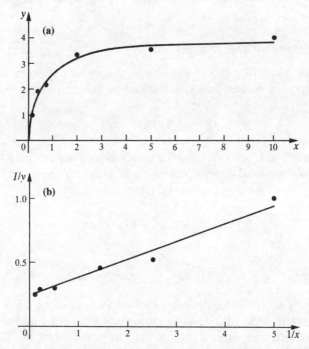

Figure 7.2 (a) A rectangular hyperbola with data points whose deviations from the curve are approximately the same. (b) The linear form of the same curve with the corresponding data points.

237

end of the ranges of values than at the other end. This means that greater weighting should be given to data points at low values of $1/x$ and $1/y$ than for points at the upper ends of these scales; and the fact that equal weighting is applied along the line results in the non-optimal fit observed. It is true that weighted regression models can be formulated, but since the proper non-linear model can be effectively employed with modern computing facilities, any 'fudging' with weighted regressions is just not worth while.

Geometrical interpretation of the least squares method of function estimation

In this section, we shall examine the theory of least squares in relation to function fitting in a little more detail. This will serve not only to give a better idea of the mechanism involved in the straightforward cases of linear regression but also to highlight important considerations when functions requiring a non-linear regression model are being fitted to data.

Linear regression: the straight line

The starting point for this discussion is the deviations sum of squares function that is minimised to give the line of best fit (eqn (7.4)):

$$S^2 = \sum (y - a - bx)^2$$

This is a function of two variables, a and b, whose surface may be plotted on a three-dimensional graph with the a- and b-axes forming the horizontal plane and S^2 being the vertical axis. Expansion gives

$$S^2 = \sum (y^2 + a^2 + b^2 x^2 - 2ay - 2bxy + 2abx)$$

i.e.

$$S^2 = \sum y^2 + a^2 n + b^2 \sum x^2 - 2a \sum y - 2b \sum xy + 2ab \sum x \qquad (7.28)$$

Remembering that a and b are the variables and the summation terms the constants, we recognise (7.28) as an example of a function of two variables of the second degree; accordingly we may write the terms of (7.28) in the order of those in (6.26):

$$S^2 = \sum y^2 - 2a \sum y + a^2 n - 2b \sum xy + b^2 \sum x^2 + 2ab \sum x \qquad (7.29)$$

238

Comparing coefficients between (6.26) and (7.29), we have

(6.26)	(7.29)
a	$\sum y^2$
b	$-2\sum y$
c	n
d	$-2\sum xy$
e	$\sum x^2$
f	$2\sum x$

The coefficients of the squares of the two variables, n and $\sum x^2$, are always positive. Furthermore, the coefficient of ab is $2\sum x$. Now if

$$4\sum{}^2x/4 - n\sum x^2$$

is negative, equation (7.29) is that of a paraboloid (see page 211). Indeed $\sum{}^2x - n\sum x^2$ must be negative because $n\sum x^2$ must always be greater than $\sum{}^2x$.

Since the coefficients of the a^2 and b^2 are positive, the paraboloid is the other way up from that shown in Figure 6.6a, with a minimum point *above* the a–b plane whose co-ordinates are precisely those values of a and b giving the straight line of best fit together with the minimum deviations sum of squares. Further, because the coefficient of ab is non-zero, the major and minor axes of the contour ellipses (Fig. 6.6b) are not parallel to the a- and b-axes. This means that the estimates of a and b are correlated (page 56) and, although this fact is of no concern in a linear regression situation, it does cause problems in non-linear cases.

A function whose parameters are such that their least squares estimates are correlated can often be reparameterised so that the functions new parameters' least squares estimates are uncorrelated. Parameters whose estimates are uncorrelated are said to be orthogonal. The straight line has an orthogonal form:

$$\hat{y} = \bar{y} + b(x - \bar{x}) \qquad (7.30)$$

where b is the gradient as before, and $\bar{y}$ is the mean of the y-values of the data. The deviations sum of squares is given by

$$S^2 = \sum\{y - \bar{y} - b(x - \bar{x})\}^2 \qquad (7.31)$$

which expands to

$$S^2 = \sum y^2 - 2\bar{y}\sum y + \bar{y}^2 n - 2b\sum(x - \bar{x})y + b^2\sum(x - \bar{x})^2 + 2\bar{y}b\sum(x - \bar{x})$$

The coefficient of the $\bar{y}b$ term is $2\sum(x - \bar{x})$, which is zero because $\sum(x - \bar{x})$

is zero. Hence the major and minor axes of the contour ellipses are parallel to the a- and b-axes, showing an absence of correlation between parameter estimates of the straight line (7.30).

Linear regression: a trigonometric function

We shall return to the function

$$y = a \cos x + b \sin x$$

Expanding (7.15), we have

$$S^2 = \sum(y^2 + a^2 \cos^2 x + b^2 \sin^2 x - 2ay \cos x - 2by \sin x + 2ab \cos x \sin x)$$

i.e.

$$S^2 = \sum y^2 + a^2 \sum \cos^2 x + b^2 \sum \sin^2 x - 2a \sum y \cos x - 2b \sum y \sin x$$
$$+ 2ab \sum \cos x \sin x$$

and, writing the terms in the order of those in (6.26), we arrive at

$$S^2 = \sum y^2 - 2a \sum y \cos x + a^2 \sum \cos^2 x - 2b \sum y \sin x + b^2 \sum \sin^2 x$$
$$+ 2ab \sum \cos x \sin x \qquad (7.32)$$

If this equation (7.32) is compared with that for the straight line situation,

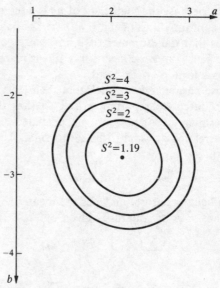

Figure 7.3 Contour ellipses on the paraboloid surface representing (7.32), using the numerical values of Example 7.1.

(7.29), it is evident that the structures of the equations are identical: only the coefficients differ, but (7.32) is still the equation of a paraboloid whose minimum is above the a–b plane and whose contour ellipses have major and minor axes not parallel with the a- and b-axes.

Figure 7.3 shows some contour ellipses associated with the paraboloid surface of (7.32) using the data of Example 7.1. The ellipses are nearly circular, showing that there is little correlation between the parameter estimates; in fact, the correlation coefficient between the estimated values of a and b is only -0.11. The derivation of such a correlation coefficient is not central to our present purpose, and we shall not pursue it.

Non-linear regression: the rectangular hyperbola

The starting point for our discussion here is the equation above (7.20) on page 232. When expanded, this gives

$$S^2 = \sum \left(y^2 + \frac{a^2 x^2}{(b+x)^2} - \frac{2axy}{b+x} \right)$$

i.e.

$$S^2 = \sum y^2 + a^2 \sum \frac{x^2}{(b+x)^2} - 2a \sum \frac{xy}{b+x} \qquad (7.33)$$

This equation is analagous to equation (7.29) for the straight line, and equation (7.32) for the trigonometric function $\hat{y} = a \cos x + b \sin x$, also a linear regression situation. The main difference between equation (7.33) and the previous two is that one of the parameters of the rectangular hyperbola, b, appears in the summation terms. This is a consequence of the rectangular hyperbola being non-linear in its parameters, and highlights one of the extra difficulties of non-linear regression. The summation terms in the fitting process do not just comprise the x- and y-values of the data in various combinations; a value of one of the function parameters must also be incorporated. This is why no explicit solutions exist to the pair of equations (7.21) and (7.22); even before the summation process is started, a value of b has to be supplied.

To draw contour curves on the surface represented by function (7.33), rearrange to

$$a^2 \sum \frac{x^2}{(b+x)^2} - 2a \sum \frac{xy}{b+x} + \left(\sum y^2 - S^2 \right) = 0 \qquad (7.34)$$

For a given value of S^2 and of b, (7.34) is a quadratic equation in a. By fixing a value of S^2 and inserting different values of b, a closed contour curve can be plotted; this has been done in Figure 7.4 for three levels of S^2 in the context of Example 7.2. The minimum point is also shown whose co-ordinates are (4.061, 0.5296, 0.1160).

241

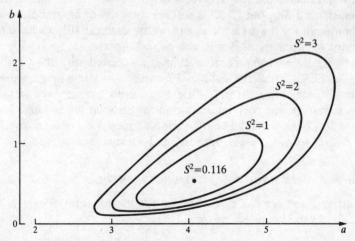

Figure 7.4 Closed contour curves on the surface representing (7.33), using the numerical values of Example 7.2.

Although the curves are closed, they are far from elliptical. Remember also that, as the curves are for increasing heights (S^2-values) of equal interval above the a–b plane, where the curves are close together a steep gradient of the surface is indicated, and *vice versa*. This means that at the top right (particularly) and bottom left ends of the curves on the graph in Figure 7.4 the gradient is relatively shallow, whereas towards the middle (around $a = 3.8$) at low values of b the gradient is very steep. This irregularity of the gradient around the minimum point may be the cause of difficulty in finding the minimum point by numerical methods such as the Newton–Raphson technique. A minimum could lie at the bottom of an elongated shallow trough, for example, making convergence very difficult. You will appreciate that the position of the starting values relative to the minimum could be of importance.

With more complicated functions that are non-linear in their parameters there could be a further problem: there may be more than one minimum point of the surface. Remember that a minimum is only a *local* phenomenon and almost certainly the S^2-value of each minimum will differ. However, in general, only one minimum will have a lower value of S^2 than any of the others and, for the purpose of function estimation, we can regard this as the true minimum. But, depending on where the starting values lie on the surface, convergence might be to another local minimum, not the true minimum. For this reason it is again evident that having the best available starting values is often crucial to success in non-linear regression.

Functions of more than one variable

There are no new principles involved if we are dealing with a function of more than one variable. In a function of two variables, a surface is being fitted to the data. Consider the plane surface

$$\hat{z} = a + bx + cy \qquad (7.35)$$

In this example, x and y are the independent variables whose values are known relatively precisely, and z is the dependent variable (usually a biological measurement in our context) whose values are subject to random variation. Thus the deviations are again in the vertical direction, in relation to a graph such as Figure 6.3a, i.e. in the z-direction only, and they can be represented as a vertical line from each datum point to the plane surface. Where the vertical line intersects the surface we have a value of z, $\hat{z}$, given by the surface itself at the x- and y-value of the datum point as calculated from (7.35). So the best fitting plane surface is given when

$$S^2 = \sum(z - \hat{z})^2$$

is minimal, i.e. when

$$S^2 = \sum(z - a - bx - cy)^2 \qquad (7.36)$$

is a minimum. The derivation of formulae for a, b, and c proceeds in the usual way and, since the function is linear in its parameters, no problems will be encountered unless the x- and y-values of the data are highly correlated.

No matter how many variables are involved, the principles of function estimation are unchanged. Any function that is linear in its parameters will generate a set of linear equations which can be explicitly solved giving parameter estimates. It is not usual to quote fomulae for the parameter estimates because the complexity of such formulae increases very rapidly with the number of parameters involved; instead, the linear equations are solved numerically for the specific set of data in hand.

Functions of more than one variable that are non-linear in their parameters are no different in principle, as regards estimation, from such functions of one variable. However, in practice, difficulties increase with an increase in the number of parameters: expert statistical advice should be sought.

Suggestions for further reading

Any textbook on statistical methods will introduce linear regression through the fitting of the straight line by least squares, and any introductory textbook on mathematical statistics will provide the theoretical background described in this chapter with, of course, much more about the statistical aspects. For books dealing solely with function estimation and related topics, the following may be noted.

A thorough but largely non-mathematical introduction to linear regression methods is provided by **Edwards, A. L.** (1984). *An introduction to linear regression and correlation*, 2nd edn. New York: W. H. Freeman.

The following book has been very popular among biologists. It gives some of the theoretical background in linear and non-linear regression as well as detailing a lot of methodology: **Draper, N. R.** and **H. Smith** (1981). *Applied regression analysis*, 2nd ed. New York: Wiley.

Much more of the theory of linear and non-linear regression is provided, but in good readable form, in the book by **Sprent, P.** (1969). *Models in regression*. London: Methuen.

Both the theory and applications of regression in growth curve methodology are described in **Causton, D. R.** and **J. C. Venus** (1981). *The biometry of plant growth*. London: Edward Arnold.

Exercises

7.1 By partially differentiating the appropriate expression for the deviations sum of squares with respect to b and $\bar{y}$, derive the formulae for b and $\bar{y}$ for the best fitting straight line of the form $\hat{y} = \bar{y} + b(x - \bar{x})$ (eqn (7.30)).

7.2 As in question 7.1, derive the formulae for a, b, and c for the best fitting curve of the form $\hat{y} = a \cos x + b \sin x + c$.

8

Differential equations I

If there could be said to exist a theoretical biology, differential equations would lie at its core. Biological processes are dynamic and so must be mathematically described by differential equations. This part of the mathematical analysis of a biological process is relatively simple, as, for example, in formulating a model to describe the growth of a microbial culture (Causton 1983, Ch. 6). The idea of a **mathematical model** or, more simply, a **model** of a biological process is very important nowadays as a means of investigating biological phenomena, and we shall discuss this topic first.

Mathematical models

A dictionary definition of a model, in the most general terms, might be 'a representation of an object'. For our purposes, the word 'representation' is the key one. The representation may be simple and crude or complex and sophisticated; indeed, one can imagine a scale of model complexity, from simple and easy to construct at one end to complex and time-consuming to make at the other end of the scale.

The above discussion may be exemplified by one class of physical models – aircraft. A model aeroplane may be so simple as to merely have the external form of a particular aeroplane and with no moving parts; such a model could 'do' nothing but stand still. A slightly more sophisticated version could have moveable landing wheels, and perhaps moveable propellor blades; further, the propellor could be driven by a twisted rubber band and, if the model were sufficiently light weight, the combination of rubber-band driven propellor and moveable wheels might cause the model aeroplane to move along a table top. More complex versions would include a more reliable and powerful motor enabling the model to actually fly a short distance. The ultimate in model aircraft is the radio-controlled model driven by a miniature internal combustion engine; this version can fly and manoeuvre in a manner comparable with a real aircraft.

Thus there is a whole spectrum of model aircraft from the easy-to-make static effigy to the complex dynamic radio-controlled version – a continuum of reality as well as complexity. In parallel with the realism and complexity aspects of this continuum, there is also a gradation of difficulty of construction and operation: the simple static model cannot be operated at all, whereas the radio-controlled model requires considerable skill to control. The two obvious skills required are firstly to manoeuvre the aeroplane in such a way that it does not crash on to the ground; and secondly, not to let it get so far away that the radio signal generated from the control box held by the operator gets too weak for effective control, for then the result would eventually also be a crash landing out of control. Indeed, it would be theoretically possible to counteract the disastrous consequence of allowing the aircraft to wander out of the range of effective radio contact by arranging a feedback mechanism. Thus, when the strength of the radio carrier wave reaches a certain (low) strength, but not low enough for control to be lost, then an automatic mechanism built into the aircraft would swing it round 180°, thus ensuring its return along a path where the strength of the carrier wave would increase. This feedback mechanism would be an extra item of model sophistication, designed to reduce the chance of inadvertent destruction of the model.

All the analogies of the model aircraft continuum − a **physical model** system − can be applied to the more abstract system of the mathematical model. First, however, let us note that mathematical models of biological processes can be divided into two groups: the **empirical** model, and the **mechanistic** model.

The empirical model is the simpler, akin to our static effigy of a model aeroplane. Such a model simply redescribes experimental data: it is a useful way of summarising the essential features of the data, but gives little or no insight into underlying processes or, to put it another way, tells us little or nothing about why we observe the result that we do. The mechanistic model, on the other hand, starts from the 'inside'. To construct a mechanistic model of an observed biological phenomenon, we consider the possible underlying mechanisms, quantify them, and put them into a mathematical form. As already stated, in biology the mathematical form of these underlying dynamic mechanisms is very often differential equations; solution of these differential equations gives us the model result, that is, how we would expect to see the progress of the biological process unfold with time. We need not restrict ourselves to modelling a biological process with respect to time; the process may be examined with respect to any other 'external' factor, e.g. temperature. The model output, or result, must then be compared with corresponding experimental data obtained under the same conditions as those subsumed in the model.

The model of the growth of a microbial culture, referred to at the beginning of this chapter, is an example of a very simple mechanistic model. Two simple assumptions are made about features controlling the growth of cells

246

in culture, each of which can be embodied in a single differential equation. The first assumption is that in a constant but unlimited environment each cell divides, on average, at a constant rate; thus $dn/dt = kn$. Secondly, as the environment becomes more limiting in respect of nutrients and accumulation of waste products, the rate of increase of cell number declines linearly with increasing cell number, i.e. $dn/dt = 1 - n/a$. These two differential equations are then combined into one, which is

$$\frac{dn}{dt} = kn\left(1 - \frac{n}{a}\right) \qquad (8.1)$$

where n is cell number at time t, and a and k are constants which have particular meaning in relation to the assumptions upon which the differential equations are based. The constants have particular numerical values according to the species of bacterium and culture conditions involved. The general solution of (8.1) is

$$n = \frac{a}{1 + be^{-kt}} \qquad (8.2)$$

and it defines the growth curve of the entire culture which can be plotted on t–n axes. A growth culture experiment is now performed with the same species of bacterium grown under the same conditions as those built into the model. Successive samples drawn from the culture, in which cell number is determined at a range of times, are then plotted as points on the same graph (see Causton 1983, Fig. 6.5); the model is finally validated by comparing the fit of the individual data points with the theoretical curve produced by the model.

The above is merely an outline of principles; it is out of place here to examine the detailed practical problems involved in model building or validation, but we close this section by noting how mathematical models are used in the furtherance of biological knowledge.

A model should be objectively **validated**; one should not rely on a graphical comparison of experimental and model results as described above. An objective statistical validatory method should decide, at a certain level of probability, whether the model is a good fit to the data or not. If the fit is bad, it means that one or more of our basic assumptions and/or numerical values of the constants are wrong. Such a 'negative' result is useful in that it forces us to question preconceived ideas; we must look for new concepts, which will give rise to different assumptions.

If the fit of a model to experimental data is good, it does not *necessarily* mean that our assumptions and parameter values are correct, only that they *may* be. A number of different assumptions giving rise to a variety of differential equations can produce similar looking solution curves to which experimental data could fit almost equally well; but only one of these different sets of assumptions could be correct. More work is needed in this case, but one use which could be tried out with the model in hand is **prediction**. In terms of our model of the growth of a microbial culture, one could use the model to predict cell number at any particular time.

Numerical solution of first-order differential equations

It is relatively easy to formulate a differential equation; it is quite another matter to solve it, that is, to obtain from the differential equation an equation not containing any derivatives. In practice, it is the exception rather than the rule to be able to solve a differential equation analytically, for example, to go from equation (8.1) to (8.2). For most differential equations arising in practice it is impossible to find a functional form of solution.

However, solutions exist to most differential equations whether or not one can express them as actual equations not containing derivatives, i.e. whether or not an *analytical* solution exists. Expression in a neat equation is not the only method of specifying a function; a detailed table comprising y-values corresponding to each of a set of x-values, from which the curve of the function may be drawn on a graph, is sufficient.

Any differential equation which has a solution can be solved *numerically*, that is, the solution can be constructed as a table of values of x and y. The process is surprisingly simple, in principle, for first-order ordinary† differential equations, considering the complexity of the methods for obtaining analytical solutions. To show the numerical method at work, and also to examine the nature of differential equations a little more deeply, we shall look at a series of simple differential equations and solve them both numerically and (where possible) analytically. This numerical procedure, which is known as **Euler's method,** is crude but is the prototype for all numerical methods of solving differential equations. In the series of equations to be examined below, the first two examples are too elementary to require even Euler's method.

The equation $dy/dx = 0$

This equation represents the gradient of a curve in the x–y plane whose gradient is everywhere zero. Now we know that the only kind of 'curve' with this property is a horizontal straight line. But *all* horizontal straight lines, having different heights above or below the x-axis (a in Fig. 8.1) are of zero gradient. Further, we know that a straight line of zero gradient and intercept a has the equation $y = a$. So $y = a$ is the *general* solution of

---

†An **ordinary differential equation** does not contain partial derivatives; a differential equation containing partial derivatives is known as a **partial differential equation.**

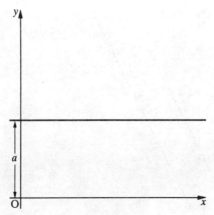

Figure 8.1 Graph of the function $y = a$.

$dy/dx = 0$. Any *particular* solution is obtained by putting a equal to some definite value, say 3; so $y = 3$ is a particular solution and its 'curve' is the horizontal straight line lying 3 units above the x-axis.

As a check on the solution, and taking a slightly more systematic approach, we are given the differential equation

$$\frac{dy}{dx} = 0 \tag{8.3}$$

with solution

$$y = a \tag{8.4}$$

We have in fact integrated (8.3) to obtain (8.4), as may be readily checked by differentiating (8.4) with respect to x. Since the derivative of a constant is zero, then our solution is correct.

The equation $dy/dx = 2$

This equation represents the gradient of a curve in the x–y plane whose gradient is everywhere 2. The gradient is thus constant, and we know that the only 'curve' having a constant gradient is a straight line. So the solution to this equation is a straight line of gradient 2, and so the relationship $y = 2x$ is a solution. The line $y = 2x$ passes through the origin (Fig. 8.2); however, there are an infinite number of other lines of gradient 2, all parallel to the line $y = 2x$, differing only in their intercepts (a in Fig. 8.2). So the general solution, corresponding to any line which could be drawn in Figure 8.2, is $y = a + 2x$; while a particular solution with, say, $a = 3$ is $y = 3 + 2x$.

Again, we should put the above discussion into a more formal, or

249

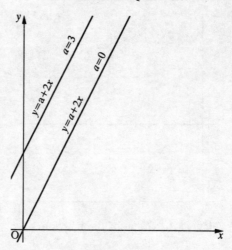

Figure 8.2 Graphs of the functions $y = 2x$ and $y = 3 + 2x$.

analytical, context. We have the differential equation

$$\frac{\mathrm{d}y}{\mathrm{d}x} = 2 \tag{8.5}$$

with solution

$$y = a + 2x \tag{8.6}$$

Equation (8.5) may be rewritten

$$\int \mathrm{d}y = 2 \int \mathrm{d}x$$

and integration yields $y = a + 2x$, where a is the combined constant of integration of the left-hand and right-hand integrals. We can see by inspection that the derivative of (8.6) is indeed (8.5).

In practice, we are often told which particular solution we need by being given the co-ordinates of a point that must lie on the solution curve: this point is called a **boundary value**. For example, if we are given, besides the differential equation $\mathrm{d}y/\mathrm{d}x = 2$, the boundary condition that $x = 2$ when $y = 7$, then substitution into the general solution, (8.6), gives $7 = a + 2(2)$, and so $a = 3$, giving the particular solution $y = 3 + 2x$.

The equation $\mathrm{d}y/\mathrm{d}x = 2x$

We first discuss this example analytically, since the gradient is no longer a constant and we are now dealing with a true curve. The differential equation is

$$\frac{dy}{dx} = 2x \tag{8.7}$$

which may be restated as

$$\int dy = 2 \int x \, dx$$

Integration yields

$$y = x^2 + c \tag{8.8}$$

where c is the combined constant of integration (see previous example). Equation (8.8) is thus the general solution of (8.7). If now the boundary condition is specified as $y = 0$ when $x = 1$, substitution in (8.8) gives $0 = 1 + c$, or $c = -1$; hence this particular solution is

$$y = x^2 - 1 \tag{8.9}$$

The graph of (8.9) is the smooth curve in Figure 8.3.

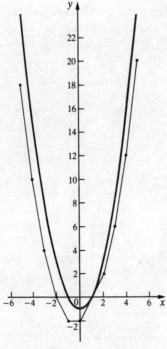

Figure 8.3 Graph of the function $y = x^2 - 1$ which is an analytical solution of the differential equation $dy/dx = 2x$, together with a numerical solution by Euler's method with interval $\Delta x(=h) = \pm 1$.

251

Now imagine that (8.7) is not integrable, and so we are unable to solve the differential equation analytically. Can we find a general formula which will generate values for the $y = f(x)$ relationship whose gradient is represented by (8.7)? For a particular solution of a differential equation this can be done.

Euler's method

In Figure 8.4a, let P be the given boundary condition (x_0, y_0); further, we know that the gradient of the curve at that point is established by the given differential equation

$$\frac{dy}{dx} \{ \text{or } y'(x_0) \} = f(x_0, y_0)$$

The term $f(x_0, y_0)$ on the right-hand side signifies that the derivative may be a function of both variables, and that it is evaluated at the point (x_0, y_0); the term $y'(x_0)$ on the left-hand side implies that the derivative is evaluated at $x = x_0$. So in Figure 8.4a, $\tan \theta = y'(x_0)$. Now consider a point Q, distant h from P in the x-direction, and where h is small; then line PR can be considered an approximation to the curve $y = f(x)$ (this line has the correct gradient at point P), and so the next point on the curve is given approximately by point R. By basic trigonometry, $\tan \theta = QR/PQ$, i.e.

$$y'(x_0) = \frac{QR}{h}$$

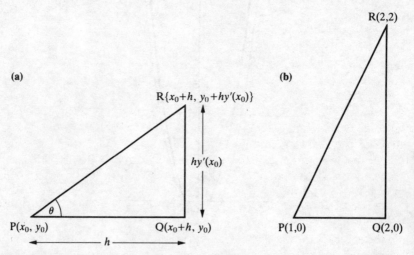

Figure 8.4 Euler's numerical method for solving a differential equation: (a) principle of the method; (b) the first interval with $\Delta x (= h) = 1$ for the equation $dy/dx = 2x$ with boundary condition $(1, 0)$.

Table 8.1 Numerical solution of $dy/dx = 2x$ with boundary point $(1, 0)$.

x	y	$y'(x)$	$x+h$	$hy'(x)$	$y+hy'(x)$
$h = 1$					
1	0	2	2	2	2
2	2	4	3	4	6
3	6	6	4	6	12
4	12	8	5	8	20
$h = -1$					
1	0	2	0	-2	-2
0	-2	0	-1	0	-2
-1	-2	-2	-2	2	0
-2	0	-4	-3	4	4
-3	4	-6	-4	6	10
-4	10	-8	-5	8	18

Hence

$$QR = hy'(x_0)$$

and so the co-ordinates of R are $\{x_0 + h, y_0 + hy'(x_0)\}$. The smaller h is, the more nearly does R lie on the true curve. Point R then forms the starting point for a further calculation. Thus in stepwise fashion one may advance from the initial point in straight-line segments generating an approximation to the solution curve.

For equation (8.7), Figure 8.4b shows the situation for the first calculation starting from the boundary condition $(1, 0)$ and using $h = 1$. Table 8.1 shows this and subsequent calculations proceeding from either side of the boundary point to $x = \pm 4$ (to work in the negative direction, simply employ a negative value of h). The resulting 'curve', consisting of straight-line segments, is also shown in Figure 8.3, and it is evident that even with the crude interval adopted ($h = \pm 1$) there is some resemblance between the curves of the analytical and numerical solutions. Nevertheless, the magnitude of the error (difference between the analytical and numerical value of y at any given x) continues to increase the further from the boundary value one goes.

Table 8.2 shows a simple computer program in the language BASIC for doing the above calculations. With this available, h may be greatly reduced and the accuracy thereby increased. Table 8.3 shows the result of doing this for the equation $dy/dx = 2x$, using different values of h. In the right-hand column are the corresponding values of the analytical solution, $y = x^2 - 1$, with which the numerically obtained results should be compared. The

Table 8.2 A simple BASIC program for solving a differential equation numerically by Euler's method.

```
 10 REM   EULER'S METHOD FOR NUMERICAL
 20 REM   SOLUTION OF A DIFFERENTIAL EQUATION
 30 REM
 40 REM   INPUT THE BOUNDARY CONDITION AND
 50 REM   DETAILS OF THE CALCULATION
 60 REM
 70 REM   XO – BOUNDARY CONDITION X-VALUE
 80 REM   YO – BOUNDARY CONDITION Y-VALUE
 90 REM   H  – INTERVAL IN X-DIRECTION
100 REM   N  – NUMBER OF DISPLAYED RESULTS REQUIRED
110 REM   M  – NUMBER OF CYCLES BETWEEN DISPLAYED
                 RESULTS
120 REM
130 INPUT XO, YO, H, N, M
140 LET X = XO
150 LET Y = YO
160 FOR I = 1 TO N
170 FOR J = 1 TO M
180 LET Y1 = 2*X
190 LET X = X + H
200 LET Y = Y + H*Y1
210 NEXT J
220 PRINT X, Y
230 NEXT I
240 END
```

Note. The above program is set up specifically to solve the equation $dy/dx = 2x$. Other equations can be solved by adjusting line 180. For example, for the equation $dy/dx = e^{-4x}$, line 180 would read LET Y1 = EXP(– 4*X).

program was run on a Honeywell 6080 computer, and the best results were obtained by setting $h = \pm 0.0001$. Use of a smaller interval gave inferior results – the accuracy of the machine was limiting. Results of greater accuracy could be obtained by using, say, a FORTRAN program with DOUBLE PRECISION.

The equation $dy/dx = 0.5y$

For an analytical solution, recast the equation in the form

$$\int \frac{dy}{y} = 0.5 \int dx$$

Table 8.3 Numerical solution of $dy/dx = 2x$ with boundary point $(1, 0)$ and different values of h.

x	$h = \pm 1$	± 0.1	± 0.01	± 0.001	± 0.0001	$\pm 0.000\,01$	$y^2 = x^2 - 1$
				y-values			
1	0	0	0	0	0	0	0
2	2	2.9	2.989 998	2.998 980	2.999 695	2.999 124	3
3	6	7.8	7.979 993	7.997 915	7.998 936	7.994 599	8
4	12	14.7	14.969 98	14.996 79	14.997 58	14.983 51	15
5	20	23.6	23.959 96	23.995 59	23.995 14	23.960 98	24
0	−2	−1.1	−1.01	−1.001 001	−1.000 114	−0.999 803 7	−1
−1	−2	−0.2	−0.019 999 4	−0.002 001	−0.000 199 5	0.000 885 6	0
−2	0	2.7	2.970 002	2.997 001	2.999 706	3.003 009	3
−3	4	7.6	7.960 004	7.996 002	7.999 604	8.005 948	8
−4	10	14.5	14.950 01	14.995 000	14.999 50	15.008 28	15
−5	18	23.4	23.940 01	23.994 02	23.999 21	24.008 08	24

which yields

$$\log_e y = 0.5x + c \qquad y > 0 \qquad (8.10)$$

Taking exponentials of both sides gives $y = e^{0.5x+c}$ or $y = e^{0.5x} e^c$. Put $a = e^c$, then

$$y = ae^{0.5x} \qquad y > 0 \qquad (8.11)$$

Because both e^c and $e^{0.5x}$ are always positive, regardless of the values of c and x, it would seem that $y > 0$ always. Thus, for example, an initial condition (another name for 'boundary condition') of $y = 3$ when $x = 0$ yields the particular solution

$$y = 3e^{0.5x}$$

This solution can also be found numerically as in the previous section (Table 8.4).

However, suppose that the initial condition was given as $y = -3$ when $x = 0$; numerically, a solution can still be obtained, as shown in Table 8.5. Evidently (8.11) is not the general solution. Where did we go wrong? The error arises in (8.10) where we have constrained y to be positive because the logarithm of only a positive number is defined. If we write (8.10) as

$$\log_e |y| = 0.5x + c \qquad y \neq 0 \qquad (8.12)$$

where $|y|$ is the modulus of y, or the absolute value of y regardless of its sign, then (8.10) is made more general. Equation (8.12) then yields

$$\pm y = e^{0.5x+c}$$

which leads to

$$y = \pm ae^{0.5x} \qquad (8.13)$$

by the same route as before. Equation (8.13) now admits all values of y, even zero if $a = 0$.

The equation $dy/dx = y/x$

Rearrangement of this equation gives

$$\int \frac{dy}{y} = \int \frac{dx}{x} \qquad x, y \neq 0$$

which, upon integration, yields

$$\log_e |y| = \log_e |x| + c \qquad x, y \neq 0$$

i.e.

$$\pm y = \exp(\log_e |x|)e^c$$

Table 8.4 Numerical solution of $dy/dx = 0.5y$ with boundary point $(0, 3)$ and different values of h.

x	$h = \pm 1$	± 0.1	± 0.01	y-values ± 0.001	± 0.0001	$\pm 0.000\,01$	$y = 3e^{0.5x}$
0	3	3	3	3	3	3	3
1	4.5	4.886 684	4.940 03	4.945 519	4.945 838	4.943 517	4.946 2
2	6.75	7.959 892	8.134 543	8.152 723	8.153 808	8.146 491	8.154 9
3	10.125	12.965 83	13.394 89	13.439 81	13.442 42	13.423 54	13.445 1
4	15.1875	21.119 96	22.056 93	22.155 61	22.161 29	22.119 33	22.167 2
−1	1.5	1.796 211	1.817 31	1.819 354	1.819 465	1.818 553	1.819 6
−2	0.75	1.075 458	1.100 872	1.103 35	1.103 489	1.102 42	1.103 6
−3	0.375	0.643 916 2	0.666 875 5	0.669 128 5	0.669 257 6	0.668 312 4	0.669 4
−4	0.1875	0.385 536 4	0.403 973 3	0.405 793 9	0.405 898	0.405 129 1	0.406 0

Table 8.5 Numerical solution of $dy/dx = 0.5y$ with boundary point $(0, -3)$ and $h = \pm 0.0001$.

x	y
0	-3
1	-4.9461
2	-8.1546
3	-13.4446
4	-22.1661
-1	-1.8196
-2	-1.1036
-3	-0.6694
-4	-0.4060

or

$$\pm y = \pm bx \qquad \text{where} \quad b = e^c$$

Only one $\pm$ sign is needed, so we may finally write as the general solution

$$y = \pm bx \tag{8.14}$$

The solution appears to be valid even at the origin, whereas the original differential equation is undefined when $x = y = 0$. Equation (8.14) represents a straight line of gradient $\pm b$, passing through the origin. Note that the boundary condition $(0, 0)$ will not define a particular solution (why?), but any other boundary condition will; for instance, if $x_0 = -1$ and $y_0 = 4$, (8.14) becomes $4 = -1b$ and so the particular solution is

$$y = -4x$$

Because the solution is a straight line, the problem is too trivial to invoke Euler's method.

The equation $dy/dx = e^{-x^2}$

In contrast to the previous examples, it is impossible to obtain an analytical solution for this equation. Hence in applying Euler's method there can be no check on the accuracy for different values of h. Accordingly, a more subjective approach to assessing the best value of h for any given computer is required. Let us seek the solution of $dy/dx = e^{-x^2}$, specifying that $y = 1$ when $x = 0$.

Table 8.6 gives calculated y-values, for various values of h, in the range

Table 8.6 Numerical solution of $dy/dx = e^{-x^2}$ with boundary point $(0, 1)$ and different values of h.

x	$h = \pm 0.1$	± 0.01	± 0.001	y-values ± 0.0001	$\pm 0.000\,01$	$\pm 0.000\,001$
0	1	1	1	1	1	1
0.1	1.1000	1.0997	1.0997	1.0997	1.0996	1.0990
0.2	1.1990	1.1976	1.1974	1.1974	1.1972	1.1959
0.3	1.2951	1.2917	1.2913	1.2912	1.2910	1.2891
0.4	1.3865	1.3804	1.3797	1.3796	1.3794	1.3768
0.5	1.4717	1.4624	1.4614	1.4613	1.4609	1.4577
0.6	1.5496	1.5367	1.5353	1.5351	1.5347	1.5309
0.7	1.6193	1.6026	1.6009	1.6007	1.6002	1.5958
0.8	1.6806	1.6600	1.6579	1.6576	1.6571	1.6521
0.9	1.7333	1.7090	1.7065	1.7062	1.7056	1.7000
1.0	1.7778	1.7500	1.7471	1.7468	1.7461	1.7400
-0.1	0.9000	0.9003	0.9003	0.9003	0.9003	0.8999
-0.2	0.8010	0.8024	0.8026	0.8026	0.8026	0.8018
-0.3	0.7049	0.7083	0.7087	0.7088	0.7087	0.7076
-0.4	0.6135	0.6196	0.6203	0.6203	0.6202	0.6188
-0.5	0.5283	0.5376	0.5386	0.5387	0.5385	0.5369
-0.6	0.4504	0.4633	0.4647	0.4648	0.4647	0.4628
-0.7	0.3807	0.3974	0.3991	0.3993	0.3991	0.3972
-0.8	0.3194	0.3400	0.3421	0.3423	0.3421	0.3402
-0.9	0.2667	0.2910	0.2935	0.2937	0.2936	0.2917
-1.0	0.2222	0.2500	0.2529	0.2531	0.2530	0.2511

$-1 \leqslant x \leqslant 1$ at $0.1x$ intervals. The results for $x = 1$ and $x = -1$, for the different values of h, are shown in Figure 8.5. At $x = -1$, the computer y-value rises with decreasing h, appearing to 'converge' to a certain y-value around $h = -10^{-4}$; but then y decreases with subsequent decrements of h. At $x = 1$, the computer y-value falls with decreasing h, again appearing to 'converge' to a certain y-value around $h = 10^{-4}$, but thereafter recommencing a decrease with successive decrements of h. Intuitively, it seems we should take $h = \pm 0.0001$ as our interval, and this value is in accord with our conclusions in the previous examples where we had an analytical solution against which our results could be assessed.

Table 8.7 shows the computed y-values for the ranges $1.1 \leqslant x \leqslant 3$ and $-3 \leqslant x \leqslant -1.1$, and from the two tables the solution curve in the range $-3 \leqslant x \leqslant 3$ can be plotted (Fig. 8.6). The curve appears to be sigmoid, with upper and lower asymptotes which may be about $y = 1.9$ and $y = 0.1$, respectively. If the curve really is sigmoid, it is symmetrical with the point of inflexion (maximum gradient) in the middle at the point $(0, 1)$.

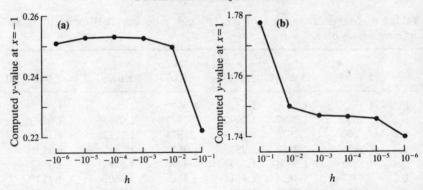

Figure 8.5 The effect of $\Delta x (= h)$ on the results of Euler's method in the solution of $dy/dx = e^{-x^2}$ with boundary value $(0, 1)$: (a) computed y-value at $x = -1$; (b) computed y-value at $x = 1$.

Although it is very easy to evaluate a numerical solution to a differential equation on a computer, every effort should be made to obtain an analytical solution if possible. In the form of an equation, all the properties of the solution can be deduced, whereas a numerical solution leaves a number of unanswered questions. For instance in the present example, what are the exact y-values of the asymptotes? Indeed, are they really asymptotes, or do the y-values start changing again when x is much bigger than 3 and/or much smaller than -3? True, the numerical solution could be extended to larger ranges of x, but one would never be certain of what lies beyond the investigated range.

Table 8.7 Numerical solution of $dy/dx = e^{-x^2}$ with boundary point $(0, 1)$ and $h = \pm 0.0001$.

x	1.1	1.2	1.3	1.4	1.5	1.6	1.7
y	1.7800	1.8067	1.8277	1.8439	1.8562	1.8652	1.8718

x	1.8	1.9	2.0	2.1	2.2	2.3	2.4
y	1.8765	1.8798	1.8820	1.8835	1.8845	1.8851	1.8855

x	2.5	2.6	2.7	2.8	2.9	3.0	
y	1.8858	1.8859	1.8860	1.8860	1.8861	1.8861	

x	-1.1	-1.2	-1.3	-1.4	-1.5	-1.6	-1.7
y	0.2199	0.1932	0.1722	0.1560	0.1437	0.1347	0.1281

x	-1.8	-1.9	-2.0	-2.1	-2.2	-2.3	-2.4
y	0.1234	0.1201	0.1179	0.1163	0.1154	0.1147	0.1143

x	-2.5	-2.6	-2.7	-2.8	-2.9	-3.0	
y	0.1141	0.1139	0.1138	0.1138	0.1137	0.1137	

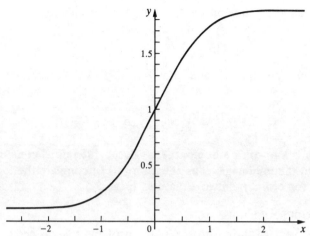

Figure 8.6 Graph of the solution of the differential equation $dy/dx = e^{-x^2}$ with boundary value $(0, 1)$ by Euler's method with $\Delta x(=h) = \pm 0.0001$.

On the other hand, for differential equations which will not yield an analytical solution, the numerical solution should not be despised. Reasonably accurate information can be obtained over ranges of x actually investigated, and in a difficult situation some knowledge is better than none.

Linear differential equations: first order

The most general form of linear differential equation is

$$f_n(x)\frac{d^n y}{dx^n} + f_{n-1}(x)\frac{d^{n-1} y}{dx_{n-1}} + \cdots + f_2(x)\frac{d^2 y}{dx^2} + f_1(x)\frac{dy}{dx} + f_0(x)y = h(x)$$

$$(8.15)$$

where $f_n(x)$, $f_{n-1}(x)$, ..., $h(x)$ are different functions of x. Thus a linear differential equation can be of any order; but nearly all equations of practical importance are of either first or second order, and our attention will be confined to these. Notice that in (8.15) only the y term and the derivatives appear linearly, the functions of x may be of any type. In general, linear equations are the simplest to deal with, and there are several well defined procedures available for their solution.

First-order equations

From (8.15) the general form of a first-order linear differential equation is

$$f_1(x)\frac{dy}{dx} + f_0(x)y = h(x) \qquad (8.16)$$

261

but on division throughout by $f_1(x)$ we can write (8.16) as

$$\frac{dy}{dx} + f(x)y = g(x) \tag{8.17}$$

where $f(x) = f_0(x)/f_1(x)$ and $g(x) = h(x)/f_1(x)$. Further, if $g(x) = 0$, (8.17) becomes

$$\frac{dy}{dx} + f(x)y = 0 \tag{8.18}$$

and (8.18) is known as a **homogeneous** linear differential equation corresponding to the **inhomogeneous** equation (8.17); compare the terminology with that for ordinary linear equations (page 22).

THE HOMOGENEOUS EQUATION

It is convenient to start with the solution of the homogeneous equation. Equation (8.18) can be arranged in the form

$$\frac{dy}{dx} = -f(x)y$$

The variables are separable, giving

$$\int \frac{dy}{y} = -\int f(x)\, dx$$

Integration yields

$$\log_e y = -\int f(x)\, dx + c \tag{8.19}$$

In these general terms we cannot do more than leave the right-hand side of (8.19) as an integral. Taking exponentials of both sides of (8.19) gives

$$y = \exp\left\{ -\int f(x)\, dx + c \right\}$$

or

$$y = \exp\left\{ -\int f(x)\, dx \right\} e^c$$

Finally, put $a = e^c$, and so we have

$$y = a \exp\left\{ -\int f(x)\, dx \right\} \tag{8.20}$$

Equation (8.20) is the general solution of (8.18) and contains one arbitrary constant, a.

Strictly, of course, the left-hand side of (8.19) should read $\log_e |y|$, and

(8.20) then has $\pm$ immediately following the $=$ sign (see page 258); but we omit these items, both here and in what follows, in order to keep the notation as simple as possible.

Example 8.1 Solve (general solution)

$$x^2 \frac{dy}{dx} + xy = 0 \qquad (8.21)$$

Divide throughout by x^2, giving

$$\frac{dy}{dx} + \frac{y}{x} = 0$$

Comparing the above with (8.18) we see that $f(x) = 1/x$ in the present example; hence substituting in (8.20) yields

$$y = a \, \exp\left(-\int x^{-1} \, dx\right)$$

which on integrating gives

$$y = a \, \exp(-\log_e x)$$

or

$$y = \frac{a}{\exp(\log_e x)}$$

and so the general solution is

$$y = a/x \qquad (8.22)$$

It is always worth while checking the solution of a differential equation. Differentiating (8.22) gives

$$\frac{dy}{dx} = -\frac{a}{x^2} \qquad (8.23)$$

and substituting (8.23) and (8.22) into (8.21) yields

$$x^2\left(-\frac{a}{x^2}\right) + x\left(\frac{a}{x}\right) = -a + a = 0$$

which is correct.

THE INHOMOGENEOUS EQUATION

Having obtained the general solution of the homogeneous equation, the inhomogeneous case now follows. First write the solution of the homogeneous equation, (8.20), as

$$y \exp\left\{\int f(x)\, dx\right\} = a \tag{8.24}$$

Now differentiate (8.24) with respect to x using the product rule (we are, in fact, reversing the solution and going back to the original equation):

$$\frac{dy}{dx} \exp\left\{\int f(x)\, dx\right\} + y\, \frac{d \exp\left\{\int f(x)\, dx\right\}}{dx} = 0$$

Differentiation of the second term is accomplished by means of the function of a function rule, giving the complete expression

$$\frac{dy}{dx} \exp\left\{\int f(x)\, dx\right\} + yf(x) \exp\left\{\int f(x)\, dx\right\} = 0$$

or

$$\left\{\frac{dy}{dx} + f(x)y\right\} \exp\left\{\int f(x)\, dx\right\} = 0 \tag{8.25}$$

Equation (8.25) is the original equation, (8.18), multiplied by

$$\exp\{\textstyle\int f(x)\, dx\}.$$

Evidently, in solving the homogeneous equation in the previous section we, in effect, multiplied by $\exp\{\int f(x)\, dx\}$ before integrating; thus $\exp\{\int f(x)\, dx\}$ is called the **integrating factor**.

Going now to the inhomogeneous equation, (8.17), multiply throughout by the integrating factor:

$$\left\{\frac{dy}{dx} + f(x)y\right\} \exp\left\{\int f(x)\, dx\right\} = g(x) \exp\left\{\int f(x)\, dx\right\} \tag{8.26}$$

The left-hand side of (8.26) is identical to the left-hand side of (8.25) which, in turn, when integrated yields the left-hand side of (8.24). Thus, integrating both sides of (8.26) gives

$$y \exp\left\{\int f(x)\, dx\right\} = \int g(x) \exp\left\{\int f(x)\, dx\right\} dx + c$$

and, on dividing throughout by $\exp\int f(x)\, dx$, we finally have

$$y = \exp\left\{-\int f(x)\, dx\right\}\left[\int g(x) \exp\left\{\int f(x)\, dx\right\} dx + c\right] \tag{8.27}$$

Equation (8.27) is the general solution of the inhomogeneous linear differential equation of the first order, (8.17).

Example 8.2 *Find the general solutions of*

$$\text{(a)} \quad \frac{dy}{dx} + \frac{y}{x} = 6x^2 \qquad \text{(b)} \quad \frac{dy}{dx} - y \tan x = \sin x$$

(a) In terms of the general equation, (8.17), $f(x) = 1/x$ and $g(x) = 6x^2$. The integrating factor is

$$\exp\left(\int x^{-1}\, dx\right) = \exp(\log_e x) = x$$

So from (8.27) we have

$$y = \frac{1}{x} \int 6x^2 x\, dx = \frac{1}{x} \int 6x^3\, dx$$

$$= \frac{1}{x}\left(\frac{6x^4}{4} + c\right)$$

i.e.

$$y = \frac{3x^3}{2} + \frac{c}{x}$$

(b) In terms of (8.17), $f(x) = -\tan x$ and $g(x) = \sin x$. The integrating factor is

$$\exp\left\{\int -(\tan x)\, dx\right\} = \exp\left\{-\int (\tan x)\, dx\right\}$$

$$= \exp\{-\log_e(\sec x)\} = \frac{1}{\sec x} = \cos x$$

(see Example 4.3, page 136).
Hence from (8.27) we have

$$y = (\sec x) \int \sin x \cos x\, dx \qquad\qquad (8.28)$$

The integrand is a product of two functions of x, and so we integrate by parts. In relationship (3.2), putting $\psi(x) = \sin x$ and $\phi(x) = \cos x$, we have

$$\int \sin x \cos x\, dx = (\sin x) \int \cos x\, dx - \int\left\{\frac{d(\sin x)}{dx} \int \cos x\, dx\right\} dx + c$$

i.e.

$$\int \sin x \cos x\, dx = \sin^2 x - \int \cos x \sin x\, dx + c$$

So

$$2 \int \sin x \cos x\, dx = \sin^2 x + c$$

265

i.e.

$$\int \sin x \cos x \, dx = \tfrac{1}{2} \sin^2 x + c$$

Substituting back into (8.28) gives

$$y = (\sec x)(\tfrac{1}{2} \sin^2 x + c)$$

The integral in (8.28) could have been more easily evaluated by substitution. Put $u = \sin x$, then $du/dx = \cos x$, and so $dx = du/\cos x$. Thus

$$\int \sin x \cos x \, dx = \int u \cos x \, \frac{du}{\cos x} = \int u \, du$$

$$= \frac{u^2}{2} + c = \tfrac{1}{2} \sin^2 x + c$$

as before.

The components of the general solution

On eliminating the square brackets in (8.27), the general solution of the inhomogeneous equation becomes

$$y = \exp\left\{-\int f(x)\, dx\right\} \int g(x) \exp\left\{\int f(x)\, dx\right\} dx + c \exp\left\{-\int f(x)\, dx\right\}$$

$$(8.29)$$

If (8.20), the general solution of the homogeneous equation, is compared with (8.29), it can be seen that the one term on the right-hand side of the former is identical with the second term on the right-hand side of the latter. Evidently the general solution of an inhomogeneous equation is similar to that of the corresponding homogeneous equation (one term is identical), but the former has an extra term. The first term on the right-hand side of (8.29) does not contain the arbitrary constant; consequently it is known as the **particular integral.** The second term, which does contain the arbitrary constant, is called the **complementary function.** Hence, the general solution of an inhomogeneous linear differential equation of the first order may be expressed in words as

general solution = particular integral + complementary function

This analysis of the general solution of an inhomogeneous equation into two parts, although not in itself important in relation to first-order equations, is of vital significance in the solution of higher-order linear equations.

FIRST ORDER LINEAR EQUATIONS

First-order equations with constant coefficients

If the function of x, $f(x)$, in (8.17) is replaced by a constant, we have

$$\frac{dy}{dx} + ay = g(x) \tag{8.30}$$

with solution

$$y = e^{-ax}\left\{ \int g(x)\, e^{ax}\, dx + c \right\} \tag{8.31}$$

The associated homogeneous equation and its solution are even simpler. What are they?

Despite their simplicity, linear equations with constant coefficients are very useful in elementary biological modelling. The following examples show the kind of processes that can be modelled.

Example 8.3   Glucose is infused into the bloodstream through a drip-feed apparatus at a constant rate of b g min^{-1}. Simultaneously, glucose is metabolised and removed from the bloodstream at a rate which is proportional to the amount of glucose present. Formulate and solve the differential equation describing these events.

Let g be the amount of glucose in grams present at time t; then from the drip

$$\frac{dg}{dt} = b$$

and from metabolism

$$\frac{dg}{dt} = -ag$$

where a is the constant of proportionality, and the minus sign denotes removal. Thus, overall

$$\frac{dg}{dt} = b - ag$$

which can be written as

$$\frac{dg}{dt} + ag = b$$

Comparing (8.30) and (8.31) we have the solution

$$g = \frac{b}{a} + ce^{-at} \tag{8.32}$$

267

where c is the arbitrary constant. If g_0 is the initial amount of glucose in the bloodstream when $t = 0$, substitution in (8.32) gives

$$g_0 = \frac{b}{a} + c$$

yielding

$$c = g_0 - \frac{b}{a}$$

so that

$$g = \frac{b}{a} + \left(g_0 - \frac{b}{a}\right) e^{-at} \tag{8.33}$$

The solution curve approaches the asymptote $g = b/a$, the quotient of the rate of infusion to the rate of metabolism.

Example 8.4 Newton's law of cooling states that the rate of decrease of temperature of a body is proportional to the difference between its temperature and that of the surroundings. Thus

$$\frac{\mathrm{d}T}{\mathrm{d}t} = -k(T - T_e)$$

where T is the temperature of the body at time t, T_e is the environmental temperature, and k is the constant of proportionality which embodies the thermal properties of the body. The solution of this equation is

$$T = T_e + be^{-kt}$$

where b is the arbitrary constant. With these simple beginnings models of animal heat balance may be constructed.

Example 8.5 A very simple model of seasonal growth is provided by the differential equation

$$\frac{\mathrm{d}n}{\mathrm{d}t} = kn \cos t$$

Here the rate of change of organism number, n, at time t is a function of both n and t. Formally, this is a homogeneous equation with a non-constant coefficient:

$$\frac{\mathrm{d}n}{\mathrm{d}t} - (k \cos t)n = 0$$

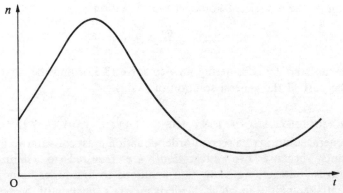

Figure 8.7 The function $n = ae^{k \sin t}$.

but for solution we note that the variables are separable, and so

$$\int \frac{dn}{n} = k \int \cos t \, dt$$

yielding

$$\log_e n = k \sin t + c$$

Taking exponentials of both sides and rearranging finally gives

$$n = ae^{k \sin t}$$

where $a = e^c$, and the curve of this relationship is shown in Figure 8.7.

Linear differential equations: second order

Second-order equations with constant coefficients

The general form of second-order linear differential equations with constant coefficients is

$$\frac{d^2 y}{dx^2} + a \frac{dy}{dx} + by = f(x) \tag{8.34}$$

because any coefficient associated with the second derivative can be divided out first.

The general solution of (8.34) may be expressed as the sum of the particular integral and the complementary function, as in the case of a first-order linear equation; likewise, the complementary function is the general

269

solution of the associated homogeneous equation

$$\frac{d^2y}{dx^2} + a\frac{dy}{dx} + by = 0 \qquad (8.35)$$

So we can start by considering how to solve (8.35), and the solution will then be part of the general solution of (8.34).

THE HOMOGENEOUS EQUATION – THE COMPLEMENTARY FUNCTION

The general solution of a second-order equation must contain two arbitrary constants, because two integrations are required to eliminate the derivatives. Further, we know that solutions to first-order homogeneous differential equations in which the derivative is a function of the dependent variable, y, involve exponentials of the independent variable, x. We are therefore led to expect that the solution of second-order linear differential equations may take the form of exponentials.

Let us assume that the solution of (8.35) may be expressed in the form

$$y = Ae^{\lambda x}$$

then

$$dy/dx = A\lambda e^{\lambda x} \quad \text{and} \quad d^2y/dx^2 = A\lambda^2 e^{\lambda x}$$

Substituting into (8.35) gives

$$A\lambda^2 e^{\lambda x} + aA\lambda e^{\lambda x} + bAe^{\lambda x} = 0$$

or

$$(\lambda^2 + a\lambda + b)Ae^{\lambda x} = 0 \qquad (8.36)$$

Now, in (8.36) either $\lambda^2 + a\lambda + b = 0$ or $Ae^{\lambda x} = 0$. But the latter is the supposed solution of the equation, and so if this were zero the solution would be trivial; so we examine the equation

$$\lambda^2 + a\lambda + b = 0 \qquad (8.37)$$

which is known as the **characteristic** or **auxiliary equation** of (8.35), and solve for λ. Let λ_1 and λ_2 be the roots of (8.37); then because of the relationship between a quadratic equation and its roots, i.e. $\lambda_1 + \lambda_2 = -a$ and $\lambda_1\lambda_2 = b$, (8.35) can be written as

$$\frac{d^2y}{dx^2} - (\lambda_1 + \lambda_2)\frac{dy}{dx} + \lambda_1\lambda_2 y = 0$$

On multiplying out the brackets, we have

$$\frac{d^2y}{dx^2} - \lambda_1\frac{dy}{dx} - \lambda_2\frac{dy}{dx} + \lambda_1\lambda_2 y = 0$$

This can be written in the form

$$\frac{d(dy/dx - \lambda_1 y)}{dx} - \lambda_2(dy/dx - \lambda_1 y) = 0$$

Put $w = \dfrac{dy}{dx} - \lambda_1 y$, then the above equation becomes

$$\frac{dw}{dx} - \lambda_2 w = 0$$

This first-order equation has the general solution

$$w = ce^{\lambda_2 x}$$

Substituting for $w = dy/dx - \lambda_1 y$ yields

$$\frac{dy}{dx} - \lambda_1 y = ce^{\lambda_2 x}$$

This is a first-order linear equation with constant coefficients, and the solution is (cf. (8.30) and (8.31))

$$y = e^{\lambda_1 x}\left\{ \int ce^{\lambda_2 x} e^{-\lambda_1 x}\, dx + c_1 \right\}$$

that is

$$y = e^{\lambda_1 x}\left\{ c \int \exp\{(\lambda_2 - \lambda_1)x\}\, dx + c_1 \right\} \qquad (8.38)$$

Integrating the right-hand side of (8.38) gives

$$y = e^{\lambda_1 x}\left\{ \frac{c}{\lambda_2 - \lambda_1} \exp\{(\lambda_2 - \lambda_1)x\} + c_1 \right\}$$

and so

$$y = c_2 e^{\lambda_2 x} + c_1 e^{\lambda_1 x} \qquad (8.39)$$

where $c_2 = c/(\lambda_2 - \lambda_1)$. Equation (8.39) is thus the general solution of (8.35), and is also the complementary function of (8.34).

If the roots of the auxiliary equation are equal, that is if $\lambda_1 = \lambda_2 = \lambda$, then (8.38) becomes

$$y = e^{\lambda x}\left\{ c_2 \int dx + c_1 \right\}$$

where c_2 has been written as the constant instead of c. Integration then yields the solution

$$y = (c_1 + c_2 x)e^{\lambda x} \qquad (8.40)$$

which again has two arbitrary constants, as it should.

It may be that the roots of (8.37) are complex; they would be complex conjugates in fact (see page 163). Let $\lambda_1 = m + in$ and $\lambda_2 = m - in$, then (8.39) becomes

$$y = c_1 \exp\{(m + in)x\} + c_2 \exp\{(m - in)x\} \qquad (8.41)$$

Now let the constants be complex conjugates also, and let $c_1 = A + iB$ and $c_2 = A - iB$; then

$$y = (A + iB) \exp\{(m + in)x\} + (A - iB) \exp\{(m - in)x\}$$

Next, eliminate the brackets and extract the constant factor e^{mx}

$$y = e^{mx}\{Ae^{inx} + iBe^{inx} + Ae^{-inx} - iBe^{-inx}\}$$

or

$$y = e^{mx}\{A(e^{inx} + e^{-inx}) + iB(e^{inx} - e^{-inx})\}$$

or (see (5.37) and (5.42))

$$y = e^{mx}(2A \cos nx + 2i^2B \sin nx)$$

and, since $i^2 = -1$

$$y = 2e^{mx}(A \cos nx - B \sin nx) \qquad (8.42)$$

Example 8.6 Find the general solutions of

(a) $\dfrac{d^2y}{dx^2} + 4\dfrac{dy}{dx} + 4y = 0$ (b) $\dfrac{d^2y}{dx^2} + \dfrac{dy}{dx} - 2y = 0$

(c) $4\dfrac{d^2y}{dx^2} + 5\dfrac{dy}{dx} + 2y = 0$

(a) The auxiliary equation is

$$\lambda^2 + 4\lambda + 4 = 0$$

and the formula for solution of a quadratic equation gives

$$\lambda = \frac{-4 \pm \sqrt{\{(4)^2 - 4(1)(4)\}}}{2}$$

giving the single root $\lambda = -2$. Hence the general solution is (eqn (8.40))

$$y = (c_1 + c_2x)e^{-2x}$$

(b) The auxiliary equation is

$$\lambda^2 + \lambda - 2 = 0$$

and the left-hand side is easily factorised (two numbers which add to give 1 and multiply to give -2):

$$(\lambda + 2)(\lambda - 1) = 0$$

giving the two roots $\lambda = -2$ and $\lambda = 1$. Thus the general solution is (eqn (8.39))

$$y = c_1 e^{-2x} + c_2 e^x$$

(c) The auxiliary equation is

$$4\lambda^2 + 5\lambda + 2 = 0$$

and using the formula for the solution of a quadratic equation gives

$$\lambda = \frac{-5 \pm \sqrt{\{(5)^2 - 4(4)(2)\}}}{8} = \frac{-5 \pm \sqrt{-7}}{8}$$

So

$$\lambda_1 = -\frac{5}{8} + i\frac{\sqrt{7}}{8} \quad \text{and} \quad \lambda_2 = -\frac{5}{8} - i\frac{\sqrt{7}}{8}$$

In terms of (8.41), $m = -\dfrac{5}{8}$ and $n = \dfrac{\sqrt{7}}{8}$, and so from (8.42)

$$y = 2e^{0.625x}\left(A \cos \frac{x\sqrt{7}}{8} - B \sin \frac{x\sqrt{7}}{8}\right)$$

or, approximately

$$y = 2e^{0.625x}(A \cos 0.3307x - B \sin 0.3307x)$$

THE INHOMOGENEOUS EQUATION: ADDING THE PARTICULAR INTEGRAL

The procedure to be described is called the **method of undetermined coefficients,** and the approach varies slightly according to the nature of $f(x)$ on the right-hand side of (8.34).

f(x) a polynomial
Here we try to find a polynomial solution as particular integral. If $f(x)$ in (8.34) is a polynomial of degree n, and $b \neq 0$, try a polynomial of degree n; if $b = 0$ but $a \neq 0$, try a polynomial of degree $n + 1$

Example 8.7   Find a particular integral of

(a) $\dfrac{d^2y}{dx^2} + \dfrac{dy}{dx} - 2y = x^2 + 4x + 7$ (b) $\dfrac{d^2y}{dx^2} + 3\dfrac{dy}{dx} = x^2 + 3x + 2$

273

DIFFERENTIAL EQUATIONS I

(a) Put $y = Ax^2 + Bx + c$; then

$$\frac{dy}{dx} = 2Ax + B \quad \text{and} \quad \frac{d^2y}{dx^2} = 2A$$

Substitution for the terms of the left-hand side of the original equation gives

$$2A + 2Ax + B - 2(Ax^2 + Bx + C) = x^2 + 4x + 7$$

i.e.

$$2A + 2Ax + B - 2Ax^2 - 2Bx - 2C = x^2 + 4x + 7$$

Now group the terms on the left-hand side as coefficients of x^2, x, and a constant:

$$-2Ax^2 + 2(A - B)x + (2A + B - 2C) = x^2 + 4x + 7$$

To make the two sides equivalent:

$$-2A = 1 \quad \text{giving} \quad A = -\tfrac{1}{2}$$

$$2(A - B) = 4 \quad \text{giving} \quad -1 - 2B = 4, \quad \text{i.e. } B = -2\tfrac{1}{2}$$

$$2A + B - 2C = 7 \quad \text{giving} \quad -1 - \tfrac{5}{2} - 2C = 7, \quad \text{i.e. } C = -5\tfrac{1}{4}$$

So the particular integral is

$$-\tfrac{1}{2}x^2 - \tfrac{5}{2}x - \tfrac{21}{4}$$

(b) Put $y = Ax^3 + Bx^2 + Cx + D$; then

$$\frac{dy}{dx} = 3Ax^2 + 2Bx + C \quad \text{and} \quad \frac{d^2y}{dx^2} = 6Ax + 2B$$

Substitution for the terms of the left-hand side of the original equation gives

$$6Ax + 2B + 3(3Ax^2 + 2Bx + C) = x^2 + 3x + 2$$

i.e.

$$6Ax + 2B + 9Ax^2 + 6Bx + 3C = x^2 + 3x + 2$$

i.e.

$$9Ax^2 + 6(A + B)x + (2B + 3C) = x^2 + 3x + 2$$

To make the two sides equivalent:

$$9A = 1 \quad \text{giving} \quad A = \tfrac{1}{9}$$

$$6(A + B) = 3, \quad \text{i.e. } A + B = \tfrac{1}{2} \quad \text{giving} \quad \tfrac{1}{9} + B = \tfrac{1}{2}, \quad \text{so } B = \tfrac{7}{18}$$

$$2B + 3C = 2, \quad \text{i.e. } \tfrac{7}{9} + 3C = 2, \quad \text{so } C = \tfrac{11}{27}$$

Further, we may take $D = 0$, so the particular integral is

$$\tfrac{1}{9}x^3 + \tfrac{7}{18}x^2 + \tfrac{11}{27}x$$

---

f(x) an exponential function
Here we try to find an exponential function as a particular integral.

Example 8.8   Find a particular integral of

(a) $\dfrac{d^2y}{dx^2} + 3\dfrac{dy}{dx} + 2y = 2e^{3x}$ *(b)* $\dfrac{d^2y}{dx^2} + 3\dfrac{dy}{dx} + 2y = e^{-x}$

(a) Put $y = Ae^{3x}$, then $dy/dx = 3Ae^{3x}$ and $d^2y/dx^2 = 9Ae^{3x}$. Substitution for the terms of the left-hand side of the original equation gives

$$9Ae^{3x} + 9Ae^{3x} + 2Ae^{3x} = 2e^{3x}$$

i.e.

$$A(9 + 9 + 2) = 2 \quad \text{or} \quad A = 0.1$$

Thus $y = 0.1e^{3x}$ is a particular integral.

(b) Put $y = Ae^{-x}$, then $dy/dx = -Ae^{-x}$ and $d^2y/dx^2 = Ae^{-x}$. Then

$$Ae^{-x} - 3Ae^{-x} + 2Ae^{-x} = e^{-x}$$

i.e.

$$A(1 - 3 + 2) = 1$$

which makes A indeterminate. The reason for this is that e^{-x} is part of the complementary function which can be shown to be $Be^{-x} + Ce^{-2x}$. This shows that in solving a linear differential equation the complementary function should always be found first.

For the particular integral, try instead $y = Axe^{-x}$, then $dy/dx = Ae^{-x}(1 - x)$ and $d^2y/dx^2 = Ae^{-x}(x - 2)$ by the product rule. Substitution into the original equation then gives

$$A(x - 2)e^{-x} + 3A(1 - x)e^{-x} + 2Axe^{-x} = e^{-x}$$

i.e.

$$A(x - 2 + 3 - 3x + 2x) = 1 \quad \text{and so } A = 1$$

Thus $y = xe^{-x}$ is a particular integral.

275

f(x) a trigonometric function

The usual way of proceeding when $f(x)$ is a trigonometric function is by replacing such functions by complex numbers. However, the method is difficult, and we shall use a much simpler method which works well for equations with constant coefficients. The method tries to find a trigonometric solution as a particular integral.

Example 8.9   Find a particular integral of

(a) $\dfrac{d^2y}{dx^2} + \dfrac{dy}{dx} + 2y = \cos x + \sin x$ (b) $3\dfrac{d^2y}{dx^2} + \dfrac{dy}{dx} + 2y = \cos x$

(c) $\dfrac{d^2y}{dx^2} + 3\dfrac{dy}{dx} + 4y = \sin 2x$

For all of these, we try the expression

$$y = A \cos nx + B \sin nx$$

from which

$$\frac{dy}{dx} = -An \sin nx + Bn \cos nx$$

and

$$\frac{d^2y}{dx^2} = -An^2 \cos nx - Bn^2 \sin nx$$

(a)   In this case n in the above expression is unity; so, substituting into the left-hand side of the original equation, we obtain

$$-A \cos x - B \sin x - A \sin x + B \cos x$$

$$+ 2A \cos x + 2B \sin x = \cos x + \sin x$$

Now equate coefficients of $\cos x$ and $\sin x$ in two separate equations,

$$(-A + B + 2A) \cos x = \cos x$$

and

$$(-B - A + 2B) \sin x = \sin x$$

i.e.

$$B + A = 1$$

and

$$B - A = 1$$

from which $B = 1$ and $A = 0$. So the particular integral is

$$y = \sin x$$

(b) Again we try $y = A \cos x + B \sin x$, and substituting on the left-hand side of the equation we have

$$-3A \cos x - 3B \sin x - A \sin x + B \cos x + 2A \cos x + 2B \sin x = \cos x$$

Equating coefficients

$$(-3A + B + 2A) \cos x = \cos x$$

$$(-3B - A + 2B) \sin x = 0$$

i.e.

$$B - A = 1$$

$$-B - A = 0$$

from which $A = -\frac{1}{2}$ and $B = \frac{1}{2}$. Hence the particular integral is

$$y = \tfrac{1}{2}(\sin x - \cos x)$$

(c) This time we try $y = A \cos 2x + B \sin 2x$, and again substituting on the left-hand side of the equation we have

$$-4A \cos 2x - 4B \sin 2x - 6A \sin 2x + 6B \cos 2x$$

$$+ 4A \cos 2x + 4B \sin 2x = \sin 2x$$

Table 8.8 A reference table indicating what should be tried for a particular integral in the solution of the equation

$$\frac{d^2 y}{dx^2} + a \frac{dy}{dx} + by = f(x)$$

$f(x)$	Trial
Polynomial of degree n (a) $b \neq 0$	polynomial of degree n
(b) $b = 0$ $a \neq 0$	polynomial of degree $n + 1$
$p e^{qx}$	$A e^{qx}$ If this fails, try $A x e^{qx}$
trigonometric function, e.g. for $\cos nx + \sin nx$	$A \cos nx + B \sin nx$

Equating coefficients

$$(-4A + 6B + 4A) \cos 2x = 0$$

$$(-4B - 6A + 4B) \sin 2x = \sin 2x$$

from which $B = 0$ and $A = -\frac{1}{6}$. Hence the particular integral is

$$y = -\frac{1}{6} \cos 2x$$

A summary of what should be tried for a particular integral for various forms of $f(x)$ is given in Table 8.8.

The general second-order linear equation

The most general form of second-order linear differential equation can be written as

$$\frac{d^2 y}{dx^2} + f(x) \frac{dy}{dx} + g(x)y = h(x) \tag{8.43}$$

since any coefficient of $d^2 y/dx^2$ can be divided out first. Only in a few cases can one find an analytical solution to (8.43); even for the simple-looking homogeneous equation

$$\frac{d^2 y}{dx^2} + xy = 0$$

no analytical solution exists. There are a few certain kinds of equation which are amenable to analytical solution by particular methods, but it takes skill to recognise these. We shall discuss here one rather more general method which may be used to solve equations of the form (8.43).

SOLVING THE SECOND-ORDER EQUATION WHEN A PARTICULAR SOLUTION OF THE ASSOCIATED HOMOGENEOUS EQUATION IS KNOWN

If any one particular solution of the homogeneous equation

$$\frac{d^2 y}{dx^2} + f(x) \frac{dy}{dx} + g(x)y = 0 \tag{8.44}$$

is known or can be determined, then the general solution of (8.43) can be obtained. Put another way, if a particular complementary function of (8.43) can be found, then the general solution can be obtained.

Let $y = u(x)$ be the known particular solution; the snag is that this might have to be obtained by a laborious process of trial and error. Introduce a further variable, z, defined by

$$y = zu(x) \tag{8.45}$$

then

$$\frac{dy}{dx} = z\,\frac{du(x)}{dx} + u(x)\,\frac{dz}{dx}$$

and

$$\frac{d^2y}{dx^2} = z\,\frac{d^2u(x)}{dx^2} + 2\,\frac{dz}{dx}\,\frac{du(x)}{dx} + \frac{d^2z}{dx^2}\,u(x)$$

Substituting for d^2y/dx^2, dy/dx, and y in (8.43), and collecting together the terms involving the different derivatives of z, we have

$$u(x)\frac{d^2z}{dx^2} + \left\{2\,\frac{du(x)}{dx} + u(x)f(x)\right\}\frac{dz}{dx}$$

$$+ \left\{\frac{d^2u(x)}{dx^2} + f(x)\,\frac{du(x)}{dx} + g(x)u(x)\right\}z = h(x) \qquad (8.46)$$

Now $u(x)$ satisfies (8.44), and so the coefficient of z in (8.46) is zero. If now we put $v = dz/dx$, (8.46) becomes

$$u(x)\,\frac{dv}{dx} + \left\{2\,\frac{du(x)}{dx} + u(x)f(x)\right\}\,v = h(x) \qquad (8.47)$$

As $u(x)$ is a known function of x – the particular solution of (8.44) we started with – so $du(x)/dx$ is also a known function of x. Thus (8.47) is a first-order equation in v and x which can be solved; the solution will contain one arbitrary constant. Because $v = dz/dx$, the solution will give dz/dx as a function of x. Solving this first-order equation then gives z as a function of x, including a second arbitrary constant; and relationship (8.45) finally gives y as a function of x which is the required solution to the original equation, (8.43).

Example 8.10   Find the general solution of

$$x^2\,\frac{d^2y}{dx^2} + (4x + 3x^2)\,\frac{dy}{dx} + (2 + 6x + 2x^2)y = x^3$$

given that a particular solution to the associated homogeneous equation is

$$y = e^{-2x}/x^2.$$

The equation can be written as

$$\frac{d^2y}{dx^2} + \frac{4x + 3x^2}{x^2}\,\frac{dy}{dx} + \frac{2 + 6x + 2x^2}{x^2}\,y = x \qquad (8.48)$$

279

In terms of (8.43),

$$f(x) = 4/x + 3, \qquad g(x) = 2/x^2 + 6/x + 2, \qquad \text{and} \qquad h(x) = x$$

Put

$$y = zu(x) = ze^{-2x}/x^2$$

then

$$\frac{dy}{dx} = -z\,\frac{2e^{-2x}(x+1)}{x^3} + \frac{dz}{dx}\frac{e^{-2x}}{x^2}$$

and

$$\frac{d^2y}{dx^2} = z\,\frac{2e^{-2x}(2x^2 + 4x + 3)}{x^4} - 4\,\frac{dz}{dx}\left\{\frac{e^{-2x}(x+1)}{x^3}\right\} + \frac{d^2z}{dx^2}\frac{e^{-2x}}{x^2}$$

On substituting into (8.48), collecting together terms in d^2z/dx^2, dz/dx, and z (as in 8.46), and eliminating the term in z (because its coefficient is zero), we have

$$\frac{e^{-2x}}{x^2}\frac{d^2z}{dx^2} + \frac{e^{-2x}}{x^3}\{4 + 3x - 4(x+1)\}\frac{dz}{dx} = x$$

which corresponds to (8.46). Simplifying further, and putting $v = dz/dx$, finally yields

$$\frac{dv}{dx} - v = x^3 e^{2x}$$

The general solution of this first-order linear equation is

$$v = e^x\left(\int x^3 e^x\,dx + c\right)$$

giving

$$v = (x^3 - 3x^2 + 6x - 6)e^{2x} + ce^x$$

where c is the constant of integration. This means that

$$\frac{dz}{dx} = (x^3 - 3x^2 + 6x - 6)e^{2x} + ce^x$$

and so

$$z = \int x^3 e^{2x}\,dx - 3\int x^2 e^{2x}\,dx + 6\int xe^{2x}\,dx - 6\int e^{2x}\,dx + c\int e^x\,dx + d$$

giving

$$z = (\tfrac{1}{2}x^3 - \tfrac{9}{4}x^2 + \tfrac{21}{4}x - \tfrac{45}{8})\,e^{2x} + ce^x + d \qquad (8.49)$$

where d is the second constant of integration. Finally, remembering that

280

SECOND ORDER LINEAR EQUATIONS

$y = zu(x)$, (8.49) needs to be multiplied by e^{-2x}/x^2, giving

$$y = \tfrac{1}{2}x - \frac{9}{4} + \frac{21}{4x} - \frac{45}{8x^2} + \frac{ce^{-x}}{x^2} + \frac{de^{-2x}}{x^2}$$

The first four terms represent the particular integral, and the last two terms are the complementary function. Therefore only the final two terms comprise the solution of the associated homogeneous equation, and $c = 0$ and $d = 1$ gives the particular solution we started with.

Uses of second-order equations

Second-order linear differential equations with constant coefficients describe oscillatory situations, and the form of an equation indicates explicitly the main kind of oscillation being described.

The physical system of a weight hanging on the end of a spring illustrates the four main situations. Free oscillation would occur if that produced by an initial displacement could continue indefinitely: the equation is

$$\frac{d^2x}{dt^2} + qx = 0$$

(where x is the displacement from the equilibrium position at time t) – a homogeneous equation with the first derivative missing. Free oscillation with damping, which of course occurs in practice, is described by the equation

$$\frac{d^2x}{dt^2} + p\frac{dx}{dt} + qx = 0$$

a homogeneous equation with the first derivative present.

If, in addition to an initial displacement of the weight, the spring's support vibrates also, we have a forced oscillation. Without damping, the equation is

$$\frac{d^2x}{dt^2} + qx = r$$

the inhomogeneous equivalent of the equation for free oscillation without damping. Finally, the situation of forced oscillation with damping has the equation

$$\frac{d^2x}{dt^2} + p\frac{dx}{dt} + qx = r$$

again, the inhomogeneous equivalent of the equation for forced oscillation without damping.

281

There are many biological phenomena which are oscillatory in nature and can be mathematically modelled by second-order differential equations. Some of these phenomena and their models are described in Jones and Sleeman (1983). A good detailed introduction to the application of second-order linear differential equations to oscillatory systems, with particular reference to the vibrating spring, is given by Curle (1972, Ch. 6).

Second-order linear equations also arise when a pair of simultaneous first-order linear equations are being solved. This topic is discussed at the beginning of Chapter 9.

The Laplace transform

The **integral transform**, $\bar{f}(p)$, of a function $f(x)$ of x in the range (a, b) is defined by

$$\bar{f}(p) = \int_a^b K(p, x)f(x)\, \mathrm{d}x \qquad (8.50)$$

where $K(p, x)$ is a function of both p and x and is called the *kernel* of the transformation. If $K(p, x) = \mathrm{e}^{-px}$, and the limits of integration are from 0 to ∞, (8.50) is known as the **Laplace transform** of $f(x)$, that is

$$\bar{f}(p) = \int_0^\infty \mathrm{e}^{-px} f(x)\, \mathrm{d}x \qquad (8.51)$$

The left-hand side of 8.51 is sometimes written as $\mathscr{L}(f)$.

The Laplace transform provides a powerful method of solving linear differential equations with constant coefficients, both ordinary and partial. The classical methods that we have discussed often require experience and judgement in providing trial forms of solution and/or the integrations can be tedious, whereas the Laplace transform method is a standard procedure for all kinds of linear equations.

To solve a linear differential equation with a given boundary condition merely requires one to evaluate the Laplace transform of each of the terms in the equation, make appropriate algebraic manipulations to obtain the dependent variable alone on the left-hand side, and finally 'back transform' to the original variables – a three-stage process.

Evaluation of Laplace transforms for different functions

First, we shall demonstrate how to find Laplace transforms for different functions by means of selected examples.

Example 8.11 *Find the Laplace transforms of the following functions of x:*

(a) c (a constant)   (b) x^n

(c) e^{ax} (d) e^{ibx}

(e) cos bx   (f) $e^{ax} \sin bx$

(a) From (8.51) we have

$$\bar{f}(p) = c \int_0^\infty e^{-px} \, dx = -\frac{c}{p} [e^{-px}]_0^\infty = \frac{c}{p} \tag{8.52}$$

(b) here $\bar{f}(p) = \int_0^\infty x^n e^{-px} \, dx = \frac{\Gamma(n+1)}{p^{n+1}} = \frac{n!}{p^{n+1}} \tag{8.53}$

(see page 94)

(c) In this case we have

$$\bar{f}(p) = \int_0^\infty e^{-px} e^{ax} \, dx = \int_0^\infty e^{-(p-a)x} \, dx$$

($p > a$, otherwise the integral does not exist)

$$= -\frac{1}{p-a} [e^{-(p-a)x}]_0^\infty$$

$$= -\frac{1}{p-a} [0-1] = \frac{1}{p-a} \tag{8.54}$$

(d) Replace a by ib in the above result:

$$\bar{f}(p) = \frac{1}{p-ib} = \frac{p+ib}{(p-ib)(p+ib)} = \frac{p+ib}{p^2 - i^2 b^2}$$

$$= \frac{p+ib}{p^2+b^2} = \frac{p}{p^2+b^2} + i \frac{b}{p^2+b^2} \tag{8.55}$$

(e) Since $e^{ibx} = \cos bx + i \sin bx$, we equate the real and imaginary part of the above result. So for $f(x) = \cos bx$,

$$\bar{f}(p) = \frac{p}{p^2+b^2} \tag{8.56}$$

and for $f(x) = \sin bx$,

$$\bar{f}(p) = \frac{b}{p^2+b^2} \tag{8.57}$$

283

(f)   This is obtained by replacing p in (8.57) by $p - a$, (8.54). So for $f(x) = e^{ax} \sin bx$,

$$\bar{f}(p) = \frac{b}{(p - a)^2 + b^2} \tag{8.58}$$

The above results and many others are given in Table 8.9. All that needs be done to use the Laplace transform to solve a linear differential equation is to utilise Table 8.9 or other extensively published tables; there is rarely any need to evaluate a Laplace transform when solving a differential equation.

Table 8.9   Laplace transforms.

	$f(x)$	$\bar{f}(p)$
i	c (constant)	c/p
ii	x^n	$n!/p^{n+1}$
iii	e^{ax}	$1/(p - a)$
iv	e^{-ax}	$1/(p + a)$
v	$x^n e^{ax}$	$n!/(p - a)^{n+1}$
vi	$x^n e^{-ax}$	$n!/(p + a)^{n+1}$
vii	$\cos bx$	$p/(p^2 + b^2)$
viii	$\sin bx$	$b/(p^2 + b^2)$
ix	$x \cos bx$	$(p^2 - b^2)/(p^2 + b^2)^2$
x	$x \sin bx$	$2bp/(p^2 + b^2)^2$
xi	$e^{ax} \cos bx$	$(p - a)/\{(p - a)^2 + b^2\}$
xii	$e^{ax} \sin bx$	$b/\{(p - a)^2 + b^2\}$
xiii	$e^{-ax} \cos bx$	$(p + a)/\{(p + a)^2 + b^2\}$
xiv	$e^{-ax} \sin bx$	$b/\{(p + a)^2 + b^2\}$
xv	$a f_1(x) + b f_2(x)$	$a \bar{f}_1(p) + b \bar{f}_2(p)$
xvi	$d\{f(x)\}/dx$	$p \bar{f}(p) - f(0)$
xvii	$d^2\{f(x)\}/dx^2$	$p^2 \bar{f}(p) - p f(0) - f'(0)$
xviii	$f(x)$, i.e. y	$\bar{f}(p)$, i.e. $\bar{y}$
xix	$(1 - e^{-ax})$	$a/\{p(p + a)\}$
xx	$(e^{-ax} + ax - 1)$	$a^2/\{p^2(p + a)\}$
xxi	$1 - \cos bx$	$b^2/\{p(p^2 + b^2)\}$
xxii	$bx - \sin bx$	$b^2/\{p^2(p^2 + b^2)\}$

Solution of linear differential equations by the Laplace transform

Solving an equation by means of the Laplace transform involves the following steps:

(1)   convert the terms of the equation into their corresponding Laplace transforms, i.e. convert $f(x)$ to $\bar{f}(p)$ in Table 8.9;

(2) rearrange to obtain y on the left-hand side and all other terms on the right-hand side;

(3) split the right-hand side into partial fractions (almost always necessary);

(4) apply inverse Laplace transforms of the right-hand side terms, i.e. convert $\bar{f}(p)$ to $f(x)$.

FIRST-ORDER EQUATIONS

Example 8.12 Solve the following first-order linear differential equations by the Laplace transform method:

(a) $\dfrac{dy}{dx} + y = e^{-x}$ $y = 3$ when $x = 0$

(b) $\dfrac{dy}{dx} + 3y = 2 \sin x$ $y = 0$ when $x = 1$

(a) From entries xvi, xviii, xv, and iv in Table 8.9, the equation becomes

$$p\bar{y} - 3 + \bar{y} = \frac{1}{p+1}$$

the 3 on the left-hand side corresponds to $f(0)$. Grouping terms gives

$$(p+1)\bar{y} = 3 + \frac{1}{p+1}$$

$$= \frac{3p+4}{p+1}$$

and so

$$\bar{y} = \frac{3p+4}{(p+1)^2}$$

Split into partial fractions (see page 173)

$$\frac{3p+4}{(p+1)^2} \equiv \frac{A}{p+1} + \frac{B}{(p+1)^2}$$

i.e.

$$3p+4 = A(p+1) + B$$

put $p = -1$,   then $-3 + 4 = B$, i.e. $B = 1$

So

$$4 + 3p = A(p+1) + 1$$

285

put $p = 0$,   then $4 = A + 1$,   i.e. $A = 3$

Hence

$$\bar{y} = \frac{3}{p+1} + \frac{1}{(p+1)^2} \qquad (8.59)$$

Now from xviii, iv, xv, and vi (8.59) becomes

$$y = 3e^{-x} + xe^{-x}$$

or

$$y = (3 + x)e^{-x}$$

(b)   From entries xvi, xviii, xv and viii in Table 8.9, the equation becomes

$$p\bar{y} - f(0) + 3\bar{y} = \frac{2}{p^2 + 1}$$

In this case we are unable to evaluate $f(0)$ because we are not told what the value of y or $f(x)$ is at $x = 0$; only the value of y at $x = 1$ is known. Proceeding as before, we have

$$(p + 3)\bar{y} = \frac{2}{p^2 + 1} + f(0)$$

$$= \frac{2 + f(0)\{p^2 + 1\}}{p^2 + 1}$$

and so

$$\bar{y} = \frac{2 + f(0)\{p^2 + 1\}}{(p + 3)(p^2 + 1)}$$

To split the right-hand side into partial fractions,

$$\frac{2 + f(0)\{p^2 + 1\}}{(p + 3)(p^2 + 1)} = \frac{A}{p + 3} + \frac{Bp + C}{p^2 + 1}$$

(see page 175).
Multiplying both sides by $(p + 3)(p^2 + 1)$ then gives

$$2 + f(0)\{p^2 + 1\} = A(p^2 + 1) + (Bp + C)(p + 3) \qquad (8.60)$$

Put $p = -3$, then

$$2 + 10f(0) = 10A, \quad \text{i.e. } A = 0.2 + f(0)$$

so (8.60) becomes

$$(Bp + C)(p + 3) + (p^2 + 1)\{0.2 + f(0)\} = 2 + f(0)\{p^2 + 1\} \qquad (8.61)$$

Now put p equal to any two convenient values in (8.61) to obtain a pair of linear equations to solve for B and C. First, put $p = 0$; then (8.61) becomes

$$3C + 0.2 + f(0) = 2 + f(0)$$

and so $C = 0.6$. Substituting for C in (8.61) and putting $p = 1$ gives

$$4(B + 0.6) + 2\{0.2 + f(0)\} = 2 + 2f(0)$$

i.e.

$$4B + 2.4 + 0.4 + 2f(0) = 2 + 2f(0)$$

and so $B = -0.2$. Thus

$$\bar{y} = \frac{0.2 + f(0)}{p + 3} + \frac{0.6 - 0.2p}{p^2 + 1}$$

$$= \frac{0.2 + f(0)}{p + 3} + \frac{0.6}{p^2 + 1} - \frac{0.2p}{p^2 + 1}$$

Then, invoking entries xviii, xv, iv, viii, and vii in Table 8.9 inversely, we have

$$y = \{0.2 + f(0)\}e^{-3x} + 0.6 \sin x - 0.2 \cos x \qquad (8.62)$$

This is the general solution, with $f(0)$ as the arbitrary constant. Putting in the boundary condition gives

$$\{0.2 + f(0)\}e^{-3} + 0.6 \sin 1 - 0.2 \cos 1 = 0 \qquad \text{(angle in radians)}$$

i.e.

$$f(0) = \frac{(0.2)(0.5403) - (0.6)(0.84147) - 0.2e^{-3}}{e^{-3}}$$

$$= -8.17$$

So the final result is

$$y = -7.97e^{-3x} + 0.6 \sin x - 0.2 \cos x \qquad (8.63)$$

Although evaluating the partial fractions is rather tedious, the application of the Laplace transform obviated the need for *two* integrations by parts.

This second example also shows that the Laplace transform can be used to obtain the general solution of a differential equation just as easily as the particular solution when $f(0)$ is known.

SECOND-ORDER EQUATIONS

Example 8.13   *Solve (general solution) the differential equation*

$$\frac{d^2 y}{dx^2} + ay = 0$$

both by the classical method and by means of the Laplace transform.

Classical method
The auxiliary equation is

$$\lambda^2 + a = 0$$

and so $\lambda = \pm i\sqrt{a}$. In terms of (8.41) $m = 0$ and $n = \sqrt{a}$, and so from (8.42)

$$y = 2(A \cos x\sqrt{a} - B \sin x\sqrt{a})$$

Laplace transform
From entries xvii, xviii, xv in Table 8.9, the equation becomes

$$p^2 \bar{y} - pf(0) - f'(0) + a\bar{y} = 0$$

i.e.

$$(p^2 + a)\bar{y} = pf(0) + f'(0)$$

or

$$\bar{y} = \frac{pf(0)}{p^2 + a} + \frac{f'(0)}{p^2 + a}$$

From entries xviii, vii, viii and xv in Table 8.9, we finally have

$$y = f(0) \cos x\sqrt{a} + \{f'(0) \sin x\sqrt{a}\}/\sqrt{a}$$

as before, where $f(0) = 2A$ and $f'(0)/\sqrt{a} = -2B$.

In this example, the Laplace transform probably involved more work than the classical method; but if the equation had been inhomogeneous, the Laplace transform would have been easier.

Example 8.14   *Solve the following second-order linear differential equations by the Laplace transform method:*

(a) $\dfrac{d^2 y}{dx^2} + \dfrac{dy}{dx} - 2y = 0$ $y = 2$ and $\dfrac{dy}{dx} = 1$ when $x = 0$

(b) $\dfrac{d^2 y}{dx^2} + \dfrac{dy}{dx} - 2y = x^2 + 4x + 7$ $y = 2$ and $\dfrac{dy}{dx} = 1$ when $x = 0$

288

(a) From Example 8.6(b), the general solution of the equation is

$$y = c_1 e^{-2x} + c_2 e^x$$

Differentiation gives

$$\frac{dy}{dx} = -2c_1 e^{-2x} + c_2 e^x$$

Substituting the boundary conditions in each of the above two equations gives

$$c_1 + c_2 = 2$$

$$-2c_1 + c_2 = 1$$

from which we obtain $c_1 = \frac{1}{3}$ and $c_2 = 1\frac{2}{3}$, and so this particular solution is

$$y = \frac{1}{3} e^{-2x} + \frac{5}{3} e^x$$

Now apply the Laplace transform to the original differential equation, using the appropriate entries from Table 8.9

$$p^2 \bar{y} - 2p - 1 + p\bar{y} - 2 - 2\bar{y} = 0$$

or

$$(p^2 + p - 2)\bar{y} = 2p + 3$$

i.e.

$$\bar{y} = \frac{2p + 3}{p^2 + p - 2}$$

Now factorise the bottom line and split into partial fractions: we have

$$\frac{2p + 3}{(p + 2)(p - 1)} = \frac{A}{p + 2} + \frac{B}{p - 1}$$

i.e.

$$2p + 3 = A(p - 1) + B(p + 2)$$

put $p = -2$,   then $-4 + 3 = -3A$, $A = \frac{1}{3}$

put $p = 1$,   then $2 + 3 = 3B$, $B = \frac{5}{3}$

So

$$\bar{y} = \frac{1}{3(p + 2)} + \frac{5}{3(p - 1)}$$

Back-transforming gives

$$y = \tfrac{1}{3} e^{-2x} + \tfrac{5}{3} e^x$$

as before.

(b)

$$\frac{d^2 y}{dx^2} + \frac{dy}{dx} - 2y = x^2 + 4x + 7$$

From Examples 8.6(b) and 8.7(a) the general solution is

$$y = c_1 e^{-2x} + c_2 e^x - \tfrac{1}{2} x^2 - \tfrac{5}{2} x - \tfrac{21}{4}$$

Differentiation gives

$$\frac{dy}{dx} = -2c_1 e^{-2x} + c_2 e^x - x - \tfrac{5}{2}$$

Substituting the boundary conditions:

$$c_1 + c_2 - \tfrac{21}{4} = 2$$

$$-2c_1 + c_2 - \tfrac{5}{2} = 1$$

from which $c_1 = \tfrac{5}{4}$ and $c_2 = 6$. So the particular solution is

$$y = \tfrac{5}{4} e^{-2x} + 6e^x - \tfrac{1}{2} x^2 - \tfrac{5}{2} x - \tfrac{21}{4}$$

Now apply the Laplace transform to the differential equation

$$p^2 \bar{y} - 2p - 1 + p\bar{y} - 2 - 2\bar{y} = \frac{2}{p^3} + \frac{4}{p^2} + \frac{7}{p}$$

i.e.

$$(p^2 + p - 2)\bar{y} = 2p + 3 + \frac{2}{p^3} + \frac{4}{p^2} + \frac{7}{p}$$

Factorising the coefficient of $\bar{y}$, $(p+2)(p-1)$, putting the right-hand side over the common denominator, p^3, and rearranging gives

$$\bar{y} = \frac{2 + 4p + 7p^2 + 3p^3 + 2p^4}{p^3 (p+2)(p-1)}$$

Now comes the really laborious part: the computation of partial fractions

$$\frac{2 + 4p + 7p^2 + 3p^3 + 2p^4}{p^3 (p+2)(p-1)} = \frac{A}{p} + \frac{B}{p^2} + \frac{C}{p^3} + \frac{D}{p+2} + \frac{E}{p-1}$$

i.e.

$$\begin{aligned} 2 + 4p + 7p^2 &+ 3p^3 + 2p^4 \\ &= Ap^2(p+2)(p-1) + Bp(p+2)(p-1) + C(p+2)(p-1) \\ &\quad + Dp^3(p-1) + Ep^3(p+2) \end{aligned}$$

put $p = 0$ from which can be calculated $C = -1$

put $p = -2$ from which is found $D = \frac{5}{4}$

put $p = 1$ and we find $E = 6$

now put $p = -1$ from which we find $A - B = -\frac{11}{4}$

finally, put $p = 2$ and we have $2A + B = -13$

From this pair of linear equations we deduce that $A = -\frac{24}{4}$ and $B = -\frac{5}{2}$; hence

$$\bar{y} = -\frac{21}{4p} - \frac{5}{2p^2} - \frac{1}{p^3} + \frac{5}{4(p+2)} + \frac{6}{p-1}$$

Back-transformation finally yields

$$y = -\frac{21}{4} - \frac{5}{2}x - \frac{1}{2}x^2 + \frac{5}{4}e^{-2x} + 6e^x$$

You would find it useful to work this example out fully for yourself by both methods (classical and Laplace) to see which you find the easier.

Suggestions for further reading

These are collated for both Chapters 8 and 9 at the end of the latter.

Exercises

8.1 Find the general solution of the following first-order linear differential equations:

(a) $\dfrac{dy}{dx} + 2y = x$

(b) $\dfrac{dy}{dx} = 2 \cos 2x - y \cot x$

(c) $\cos x \dfrac{dy}{dx} + y \sin x = 1$

8.2 By finding the complementary function and particular integral, or by using the Laplace transform, obtain the general solutions of the following second-order linear differential equations:

(a) $\dfrac{d^2y}{dx^2} + 2\dfrac{dy}{dx} + 5y = 1 + x^2$

(b) $\dfrac{d^2y}{dx^2} + 3\dfrac{dy}{dx} + 2y = e^{-x} + 2e^{-2x}$

(c) $\dfrac{d^2y}{dx^2} + 4y = \cos 2x$

8.3 Use the Laplace transform to obtain the general solution of

$$\frac{d^2y}{dx^2} - 7\frac{dy}{dx} + 12y = 7e^{-3x}$$

8.4 Find the general solution of

$$x^2\frac{d^2y}{dx^2} - 4x\frac{dy}{dx} + 6y = x^5 \sin x$$

given that a particular solution to the associated homogeneous equation is $y = x^3 + x^2$.

9

Differential equations II

In this final chapter, a diversity of topics in the subject of differential equations will be introduced. Firstly, simultaneous equations, in which we consider the simplest situation, namely, a pair of first-order linear differential equations in two unknowns; such considerations then lead on to models of interacting species. Secondly, in a short section on non-linear equations, we deal with a few standard kinds that are solvable analytically. Thirdly, there is a *very* brief introduction to partial differential equations, and finally we consider a topic of great importance in modern mathematical biology – stability analysis.

Simultaneous differential equations

In some situations there may be more than one variable which is dependent on an independent variable and whose relationships are defined by a set of differential equations. Obvious biological examples are: predator–prey, host–parasite, and competition of coexisting plant species; situations where the rates of change of, say, the host species and the parasite species, at time t are given by a model involving a pair of simultaneous differential equations. Such models of interacting species do not, however, give the simplest set of equations; so we shall return to this subject later.

Linear equations

The simplest set of simultaneous differential equations is first-order linear, with constant coefficients, of the form

$$a_1\frac{dx}{dt} + b_1\frac{dy}{dt} + c_1 x + d_1 y = f_1(t) \tag{9.1}$$

$$a_2\frac{dx}{dt} + b_2\frac{dy}{dt} + c_2 x + d_2 y = f_2(t) \tag{9.2}$$

where the a_i, b_i, c_i, and d_i are constants, and x and y are the unknowns to be found as functions of t.

Multiply (9.1) by b_2 and (9.2) by b_1, giving

$$a_1b_2 \frac{dx}{dt} + b_1b_2 \frac{dy}{dt} + b_2c_1x + b_2d_1y = b_2f_1(t)$$

$$a_2b_1 \frac{dx}{dt} + b_1b_2 \frac{dy}{dt} + b_1c_2x + b_1d_2y = b_1f_2(t)$$

Subtraction gives

$$(a_1b_2 - a_2b_1) \frac{dx}{dt} + (b_2c_1 - b_1c_2)x + (b_2d_1 - b_1d_2)y = b_2f_1(t) - b_1f_2(t)$$

$$(9.3)$$

There are now various possibilities.

(1) $(a_1b_2 - a_2b_1) \neq 0,$ $(b_2d_1 - b_1d_2) \neq 0$

Rearrange (9.3) to give y in terms of x and dx/dt:

$$y = \frac{b_2f_1(t) - b_1f_2(t)}{b_2d_1 - b_1d_2} - \frac{a_1b_2 - a_2b_1}{b_2d_1 - b_1d_2} \frac{dx}{dt} - \frac{b_2c_1 - b_1c_2}{b_2d_1 - b_1d_2} x \quad (9.4)$$

and differentiate with respect to t to give

$$\frac{dy}{dt} = \frac{b_2f_1'(t) - b_1f_2'(t)}{b_2d_1 - b_1d_2} - \frac{a_1b_2 - a_2b_1}{b_2d_1 - b_1d_2} \frac{d^2x}{dt^2} - \frac{b_2c_1 - b_1c_2}{b_2d_1 - b_1d_2} \frac{dx}{dt} \quad (9.5)$$

Finally, substitution of (9.4) and (9.5) into either (9.1) or (9.2) gives a second-order linear equation in x with constant coefficients. This can be solved by the methods given in Chapter 8, and the solution will contain two arbitrary constants. Having found x as a function of t (and, by differentiation, dx/dt as a function of t), y as a function of t can be found by substituting into (9.4).

(2) $(a_1b_2 - a_2b_1) \neq 0,$ $(b_2d_1 - b_1d_2) = 0$

Equation (9.3) will now be

$$(a_1b_2 - a_2b_1) \frac{dx}{dt} + (b_2c_1 - b_1c_2)x = b_2f_1(t) - b_1f_2(t)$$

a first-order equation whose general solution will contain one arbitrary constant. Once x is known as a function of t, and by differentiation dx/dt as a function of t, substitution into either (9.1) or (9.2) gives a first-order linear equation in y whose solution introduces a second arbitrary constant.

(3) $(a_1b_2 - a_2b_1) = 0,$ $(b_2d_1 - b_1d_2) \neq 0$

Equation (9.3) is now

$$(b_2c_1 - b_1c_2)x + (b_2d_1 - b_1d_2)y = b_2f_1(t) - b_1f_2(t)$$

Rearrangement gives

$$y = \frac{b_2 f_1(t) - b_1 f_2(t)}{b_2 d_1 - b_1 d_2} - \frac{b_2 c_1 - b_1 c_2}{b_2 d_1 - b_1 d_2} x \tag{9.6}$$

and differentiation with respect to t yields

$$\frac{dy}{dt} = \frac{b_2 f_1'(t) - b_1 f_2'(t)}{b_2 d_1 - b_1 d_2} - \frac{b_2 c_1 - b_1 c_2}{b_2 d_1 - b_1 d_2} \frac{dx}{dt} \tag{9.7}$$

Substitution of (9.6) and (9.7) into either (9.1) or (9.2) gives a first-order equation in x whose general solution contains one arbitrary constant. With x known as a function of t, substitution into (9.6) will yield y as a function of t, but with no additional arbitrary constant.

(4) $(a_1 b_2 - a_2 b_1) = 0,$ $(b_2 d_1 - b_1 d_2) = 0,$ $(b_2 c_1 - b_1 c_2) \neq 0$
Equation (9.3) is now

$$(b_2 c_1 - b_1 c_2)x = b_2 f_1(t) - b_1 f_2(t)$$

which gives x immediately in terms of t. Substitution in either (9.1) or (9.2) then gives y with one arbitrary constant.

(5) $(a_1 b_2 - a_2 b_1) = 0,$ $(b_2 d_1 - b_1 d_2) = 0,$ $(b_2 d_1 - b_1 c_2) = 0$
Now (9.3) becomes

$$b_2 f_1(t) - b_1 f_2(t) = 0$$

which implies either that (9.1) and (9.2) are inconsistent or that one is simply a multiple of the other.

One useful summary of the above is that if the coefficient of dx/dt in (9.3) is not zero the general solution of (9.1) and (9.2) contains two arbitrary constants, whereas if dx/dt vanishes, the general solution contains only one arbitrary constant.

Example 9.1 Solve the pair of equations

$$\frac{dx}{dt} - \frac{dy}{dt} - 4x + 4y = 0 \tag{9.8}$$

$$-2\frac{dx}{dt} + 3\frac{dy}{dt} + 3x - 6y = e^{-3t} \tag{9.9}$$

Multiply (9.8) by 3:

$$3\frac{dx}{dt} - 3\frac{dy}{dt} - 12x + 12y = 0$$

$$-2\frac{dx}{dt} + 3\frac{dy}{dt} + 3x - 6y = e^{-3t}$$

Because the coefficients of dy/dt are of opposite sign, we add the two equations rather than subtract them, giving

$$\frac{dx}{dt} - 9x + 6y = e^{-3t} \tag{9.10}$$

All the coefficients of (9.10) are non-zero, so we have case (1) above.
From (9.10), we have

$$y = \tfrac{1}{6}e^{-3t} - \tfrac{1}{6}\frac{dx}{dt} + \tfrac{3}{2}x \tag{9.11}$$

and

$$\frac{dy}{dt} = -\tfrac{1}{2}e^{-3t} - \tfrac{1}{6}\frac{d^2x}{dt^2} + \tfrac{3}{2}\frac{dx}{dt} \tag{9.12}$$

So, substituting into (9.8) and collecting together terms in d^2x/dt^2, dx/dt, x, and e^{-3t}, we obtain

$$\tfrac{1}{6}\frac{d^2x}{dt^2} - \tfrac{7}{6}\frac{dx}{dt} + 2x + \tfrac{7}{6}e^{-3t} = 0$$

Multiplication throughout by 6, and a slight rearrangement, gives a standard second-order linear equation with constant coefficients:

$$\frac{d^2x}{dt^2} - 7\frac{dx}{dt} + 12x = -7e^{-3t} \tag{9.13}$$

The solution is (see Exercise 8.3 at end of Chapter 8)

$$x = (c_1 - 1)e^{4t} + (c_2 + 1)e^{3t} \tag{9.14}$$

from which

$$\frac{dx}{dt} = 4(c_1 - 1)e^{4t} + 3(c_2 + 1)e^{3t} \tag{9.15}$$

Finally, substitution of (9.14) and (9.15) into (9.11) and the collection together of terms in e^{-3t}, e^{3t}, e^{4t}, c_1e^{4t}, and C_2e^{3t} gives

$$y = \tfrac{1}{6}e^{-3t} + \tfrac{5}{6}(c_1 - 1)e^{4t} + (c_2 + 1)e^{3t} \tag{9.16}$$

If only one derivative occurs in each equation, for example dy/dt in (9.1) and dx/dt in (9.2), then a slightly different initial approach is needed.

Example 9.2   Solve the pair of equations

$$\frac{dy}{dt} + 4x - 4y = 0 \tag{9.17}$$

$$2\frac{dx}{dt} - 3x + 6y = e^{-3t} \tag{9.18}$$

Differentiate (9.17) with respect to t:

$$\frac{d^2y}{dt^2} + 4\frac{dx}{dt} - 4\frac{dy}{dt} = 0$$

i.e.

$$\frac{dx}{dt} = \frac{dy}{dt} - \tfrac{1}{4}\frac{d^2y}{dt^2}$$

Substitute for dx/dt in (9.18), giving

$$2\frac{dy}{dt} - \tfrac{1}{2}\frac{d^2y}{dt^2} - 3x + 6y = e^{-3t} \qquad (9.19)$$

From (9.17), we have

$$x = y - \tfrac{1}{4}\frac{dy}{dt}$$

Substitute for x in (9.19); collect together terms in d^2y/dt^2, dy/dt, and y, and multiply throughout by -2:

$$\frac{d^2y}{dt^2} - \frac{11}{2}\frac{dy}{dt} - 6y = -2e^{-3t} \qquad (9.20)$$

As before, we have a second-order equation in y. After solving (9.20) for y, and differentiating the result, substitution into (9.17) will give the result for x.

A thorough introduction to systems of differential equations is given by Jones and Sleeman (1983). Also, some simple applications of linear systems of differential equations to species interactions (competition, predator–prey, etc.) are presented in Grossman and Turner (1974).

The Lotka–Volterra model of host–parasite relationships

The Lotka–Volterra model was one of the earliest models of species interaction. Denoting the number of parasites in the system at time t by p, and the number of hosts by h, the model is given by

$$\frac{dh}{dt} = (a_1 - b_1 p)h \qquad (9.21)$$

$$\frac{dp}{dt} = (-a_2 + b_2 h)p \qquad (9.22)$$

The meanings of these equations is quite clear. Equation (9.21) shows that the rate of change of host number, at time t, is both directly proportional to the existing host number and linearly related to parasite number. The linear function has a negative gradient consistent with the idea that the more parasites there are the lower the rate of increase of the host species. Equation (9.22) shows that the rate of change of parasite number, at time t, is both directly proportional to the existing number of parasites and linearly related to the host number. The linear function has a positive gradient consistent with the idea that the more hosts there are the higher the rate of increase of the parasites.

Combining the two equations (9.21) and (9.22), in a similar manner to that shown in Example 9.2, leads to a non-linear second-order differential equation describing either host or parasite rate of change, and no analytical solution is possible. For example, the differential equation describing host rate of change is

$$\frac{d^2h}{dt^2} - \frac{1}{h}\left(\frac{dh}{dt}\right)^2 + \left(\frac{a_2}{h} - b_2\right)\frac{dh}{dt} + a_1(b_2h - a_2) = 0$$

However, the consequence of the model for a host–parasite relationship can be deduced quite simply in another way. Divide (9.21) by (9.22), giving

$$\frac{dh}{dp} = \frac{(a_1 - b_1p)h}{(-a_2 + b_2h)p}$$

The variables are separable, and so

$$(-a_2 + b_2h)\frac{dh}{h} = (a_1 - b_1p)\frac{dp}{p}$$

or

$$-a_2\frac{dh}{h} + b_2\,dh - a_1\frac{dp}{p} + b_1\,dp = 0$$

Integration gives

$$-a_2\log_e h + b_2h - a_1\log_e p + b_1p = c \tag{9.23}$$

Equation (9.23) represents a closed curve in the h–p plane. In fact, for given values of a_1, b_1, a_2, b_2, (9.23) gives an infinite number of curves, each one corresponding to a particular value of the constant c (Fig. 9.1). Each curve gives a concise description of the host–parasite relationship, in terms of the numbers of each at any one time, by following the curve round in an anticlockwise direction from any starting point which is defined by a host number, h, and a parasite number, p. It can be seen that host numbers and parasite numbers oscillate indefinitely with parasite number changes lagging behind those of the host.

In equation (9.23) and Figure 9.1 time does not appear explictly but is, or course, implicit. Changes in host and parasite number do take place in time. The plane depicted in Figure 9.1 is known as a **phase plane**, and the

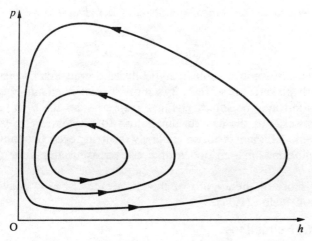

Figure 9.1 Graphs of the equation $-0.5 \log_e h + 0.02\,h - \log_e p + 0.1\,p = c$ for three different values of c.

curves with their arrows are called **trajectories**. Phase planes are very useful for demonstrating changes in two variables which are both dependent on each other (the hosts and parasites in this example) and on a third variable (time in this example), and we shall return to consider an important aspect of phase planes at the end of this chapter. Although we are dealing with three variables, the situation is not the same as those described in Chapter 6, where there were two independent and one dependent variables. Here one could say that there is only one independent variable (time) and two dependent variables (species numbers), each of which is dependent on each other as well as on the independent variable.

The model of Leslie and Gower

This model is very similar to the Lotka–Volterra model, and is defined by the pair of equations

$$\frac{dh}{dt} = (a_1 - b_1 p)h \tag{9.24}$$

$$\frac{dp}{dt} = (a_2 - b_2 p/h)p \tag{9.25}$$

The rationale of the p/h term is that when parasites are numerous and hosts are few, p/h is large, and growth rate of the parasite population slows down. Conversely, when the number of hosts is ample for the parasites present, p/h is small, and there is only a slight restriction on parasite popula-

299

tion growth rate. Division of (9.25) by (9.24) gives

$$\frac{dp}{dh} = \frac{(a_2h - b_2p)p}{(a_1 - b_1p)h^2}$$

This equation is not solvable analytically, and even Euler's method is not sufficiently powerful to solve it. By a more sophisticated numerical method, the equation may be solved giving a trajectory on the h–p plane which spirals inwards to the equilibrium point. So, in contrast to the Lotka–Volterra model, there is not an endlessly recurring cycle but convergence to the equilibrium point in cycles of ever-decreasing amplitude (see Pielou 1977).

Many more of these kinds of models are described in Maynard-Smith (1974) and Pielou (1977).

Non-linear equations

Most non-linear differential equations arising in practice cannot be solved analytically, but there are a few exceptions which will be briefly surveyed in the following paragraphs.

Variables separable

Differential equations of this kind are the simplest to deal with, and were introduced in Causton (1983). Both linear and non-linear examples exist.

Homogeneous equations

The so-called homogeneous differential equation takes the form

$$\frac{dy}{dx} = f\left(\frac{y}{x}\right) \tag{9.26}$$

To solve (9.26), make the substitution

$$y = vx \tag{9.27}$$

where v is a new variable. Differentiating (9.27) by the product rule gives

$$\frac{dy}{dx} = v + x\frac{dv}{dx}$$

and substituting in (9.26) yields

$$v + x\frac{dv}{dx} = f(v)$$

or

$$x\frac{dv}{dx} = f(v) - v \tag{9.28}$$

The variables in (9.28) are separable, and so may be solved for v; finally, y is obtained from (9.27).

Example 9.3   Solve the equation

$$\frac{dy}{dx} = \frac{x + 3y}{3x + y}$$

Divide numerator and denominator of the right-hand side by x, giving

$$\frac{dy}{dx} = \frac{1 + 3(y/x)}{3 + (y/x)}$$

With substitution (9.27) this becomes

$$v + x\frac{dv}{dx} = \frac{1 + 3v}{3 + v}$$

i.e.

$$x\frac{dv}{dx} = \frac{1 + 3v}{3 + v} - v$$

or

$$x\frac{dv}{dx} = \frac{1 + 3v - v(3 + v)}{3 + v} = \frac{1 + 3v - 3v - v^2}{3 + v}$$

and so

$$x\frac{dv}{dx} = \frac{1 - v^2}{3 + v}$$

Invert both sides:

$$\frac{dx}{x} = \frac{(3 + v)\,dv}{1 - v^2}$$

The right-hand side can be integrated by partial fractions; we have

$$\frac{3 + v}{1 - v^2} \equiv \frac{3 + v}{(1 - v)(1 + v)} \equiv \frac{A}{1 - v} + \frac{B}{1 + v}$$

i.e.

$$3 + v \equiv A(1 + v) + B(1 - v)$$

$$\text{put } v = 1, \qquad \text{then } 4 = 2A \quad \text{or} \quad A = 2$$

$$\text{put } v = -1, \qquad \text{then } 2 = 2B \quad \text{or} \quad B = 1$$

Thus

$$\frac{3+v}{1-v^2} \equiv \frac{2}{1-v} + \frac{1}{1+v}$$

Hence

$$\log_e x = -2 \log_e(1-v) + \log_e(1+v) + c$$

Taking exponentials gives

$$x = \frac{a(1+v)}{(1-v)^2}$$

where $a = e^c$. Substituting for v:

$$x(1 - y/x)^2 = a(1 + y/x)$$

i.e.

$$(x-y)^2 = a(x+y)$$

The Bernoulli equation

The Bernoulli equation takes the form

$$\frac{dy}{dx} + f(x)y = g(x)y^n \tag{9.29}$$

which is very similar to a linear equation except for the term y^n. Divide throughout by y^n, yielding

$$\frac{1}{y^n}\frac{dy}{dx} + f(x)\frac{1}{y^{n-1}} = g(x) \tag{9.30}$$

Make the substitution

$$v = \frac{1}{y^{n-1}} = y^{1-n} \tag{9.31}$$

and so

$$\frac{dv}{dx} = (1-n)y^{-n}\frac{dy}{dx} \tag{9.32}$$

Substituting (9.31) and (9.32) into (9.30) gives

$$\frac{1}{1-n}\frac{dv}{dx} + f(x)v = g(x)$$

or

$$\frac{dv}{dx} + (1 - n)f(x)v = (1 - n)g(x) \qquad (9.33)$$

which is a linear differential equation in v.

Example 9.4 *Solve the equation*

$$\frac{dy}{dx} = xy(1 + y)$$

On multiplying out, we have

$$\frac{dy}{dx} = xy + xy^2$$

or

$$\frac{dy}{dx} - xy = xy^2$$

In the notation of (9.29), $f(x) = -x$, $g(x) = x$, and $n = 2$. Hence, equation (9.33) is

$$\frac{dv}{dx} + xv = -x$$

Applying the formula (8.27), where now $f(x) = x$ and $g(x) = -x$, we have

$$v = \exp\left(-\int x \, dx\right)\left\{\int -x \exp\left(\int x \, dx\right) dx + c\right\}$$

$$= e^{-\frac{1}{2}x^2}\left\{\int -x e^{\frac{1}{2}x^2} \, dx + c\right\}$$

$$= e^{-\frac{1}{2}x^2}(-e^{\frac{1}{2}x^2} + c)$$

So

$$v = c e^{-\frac{1}{2}x^2} - 1$$

Hence from (9.31)

$$y^{-1} = c e^{-\frac{1}{2}x^2} - 1$$

or

$$y = (c e^{-\frac{1}{2}x^2} - 1)^{-1}$$

The Riccati equation

The differential equation

$$\frac{dy}{dx} + f(x)y + g(x)y^2 = h(x) \tag{9.34}$$

is called the Riccati equation. It includes forms previously examined: thus if $g(x) = 0$ it becomes a linear equation, whereas if $h(x) = 0$ we have a Bernoulli equation. If neither $g(x)$ nor $h(x)$ are zero, the equation is usually only solvable numerically. However, if any one particular solution can be found then the general solution can be formulated. This situation is similar to that of the general second-order linear equation (page 278).

Let $y = u(x)$ be the known particular solution; then define

$$y = u(x) + \frac{1}{v} \tag{9.35}$$

where v is a function of x. Differentiating (9.35) gives

$$\frac{dy}{dx} = \frac{du(x)}{dx} - \frac{1}{v^2}\frac{dv}{dx} \tag{9.36}$$

Substitution of (9.35) and (9.36) into (9.34) yields

$$\frac{du(x)}{dx} - \frac{1}{v^2}\frac{dv}{dx} + f(x)\left\{u(x) + \frac{1}{v}\right\} + g(x)\left\{u^2(x) + 2\frac{u(x)}{v} + \frac{1}{v^2}\right\} = h(x) \tag{9.37}$$

But since $y = u(x)$ is a solution of (9.34), then

$$\frac{du(x)}{dx} + f(x)u(x) + g(x)u^2(x) = h(x) \tag{9.38}$$

Now subtract (9.38) from (9.37):

$$-\frac{1}{v^2}\frac{dv}{dx} + f(x)\frac{1}{v} + 2g(x)u(x)\frac{1}{v} + g(x)\frac{1}{v^2} = 0$$

i.e.

$$-\frac{1}{v^2}\frac{dv}{dx} + \{f(x) + 2g(x)u(x)\}\frac{1}{v} + g(x)\frac{1}{v^2} = 0$$

or

$$\frac{dv}{dx} - \{f(x) + 2g(x)u(x)\}v = g(x) \tag{9.39}$$

Equation (9.39) is linear in v; when solved, use of (9.35) gives the final general solution of (9.34).

Example 9.5 Solve the equation

$$\frac{dy}{dx} + 4y - xy^2 = \frac{3}{x} - \frac{1}{x^2}$$

This equation is of the form of (9.34) with $f(x) = 4$, $g(x) = -x$, and $h(x) = 3/x - 1/x^2$.

Now $y = 1/x$ is a particular solution of this equation, which you can check by substitution; so $u(x) = 1/x$. Substituting in (9.39), we have

$$\frac{dv}{dx} - \left\{4 + 2(-x)\left(\frac{1}{x}\right)\right\}v = -x$$

i.e.

$$\frac{dv}{dx} - 2v = -x$$

Applying the formula (8.27), where now $f(x) = -2$ and $g(x) = -x$, we get

$$v = \exp\left(-\int -2 \, dx\right)\left\{\int -x \exp\left(\int -2 \, dx\right) dx + c\right\}$$

or

$$v = e^{2x}\left(c - \int xe^{-2x} \, dx\right)$$

Integration by parts gives the result

$$\int xe^{-2x} \, dx = -\tfrac{1}{2}e^{-2x}(x + \tfrac{1}{2})$$

and so

$$v = e^{2x}\{c + \tfrac{1}{2}e^{-2x}(x + \tfrac{1}{2})\}$$

or, more simply,

$$v = ce^{2x} + \tfrac{1}{2}(x + \tfrac{1}{2})$$

Finally, applying (9.35), we have

$$y = \frac{1}{x} + \frac{1}{ce^{2x} + \tfrac{1}{2}(x + \tfrac{1}{2})}$$

which is the general solution required.

Partial differential equations

In a situation where a variable is dependent on more than one independent variable, say $z = f(x, y)$, partial differential equations may arise. A **partial**

differential equation is an equation containing partial derivatives; for example

$$\frac{\partial z}{\partial y} + 2y \frac{\partial z}{\partial x} - z = 0 \tag{9.40}$$

and

$$(x^2 + y^2) \frac{\partial^2 z}{\partial y^2} - x \frac{\partial^2 z}{\partial x^2} + y \frac{\partial z}{\partial x} = \sin x + \cos y \tag{9.41}$$

exemplify a first-order and second-order partial differential equation, respectively.

As in the case of ordinary differential equations, partial differential equations can be divided into linear and non-linear types. A **linear partial differential equation** in three variables, where z is the dependent variable, has terms in z and its partial derivatives to the first degree only, and products of z and its derivatives are absent. A linear partial differential equation is homogeneous if each term contains either the dependent variable or one of its derivatives and non-homogeneous if it contains terms not comprising either the dependent variable or one of its derivatives. Thus equation (9.40) is homogeneous whereas (9.41) is non-homogeneous.

Except for certain types of linear equation, it is not possible to obtain analytical general solutions of partial differential equations; numerical methods have to be resorted to. On the other hand, the linear equations which can be solved analytically have important applications in the physical sciences. However, the general solution of a partial differential equation is of little practical use since it has to be made to satisfy boundary conditions. This is much more difficult to accomplish for partial differential equations than for ordinary differential equations for the following reason.

As we have seen, the general solution of an ordinary differential equation involves an arbitrary constant. The general solution of a partial differential equation, on the other hand, involves an arbitrary function. To see this, consider the relationship

$$z = yf(x) \tag{9.42}$$

Differentiation with respect to y gives

$$\frac{\partial z}{\partial y} = f(x)$$

and elimination of $f(x)$ between these two equations yields

$$y \frac{\partial z}{\partial y} = z \tag{9.43}$$

which is a first-order partial differential equation whose general solution is given by (9.42). The important point is that the solution of (9.43) contains an arbitrary function, $f(x)$ in this example. Now considerable choice is available for defining the arbitrary function, and so for this and other reasons, methods of solution which 'build' the solution of a partial differential equation around the boundary conditions are used in practice.

From these few remarks, it is clear that the subject of partial differential equations is a complicated one, and we shall pursue it no further here. A good introduction, though in the context of physical science, is given by Stephenson (1970); Jones and Sleeman (1983) provide a less extensive introduction to partial differential equations, but in a biological setting.

Stability analysis

In the two-species models of population dynamics presented earlier in this chapter (pages 297–300), the idea of the phase plane was introduced. The phase plane shows the relationship between two interacting species in the form of a trajectory. The function defining the trajectory is the solution of the differential equations defining the model, and, usually, realistic models comprise differential equations which are not amenable to analytical solution.

An important question concerning the interaction of species is the behaviour of the trajectory in the neighbourhood of the equilibrium (see page 300). The function defining the trajectory will, of course, give this information, but solving the original system of differential equations is not easy, whether analytically or numerically. The method of stability analysis allows us to investigate the nature of the equilibrium directly from the set of differential equations describing the model.

One-variable system

To fix ideas, we shall start with a single species population whose rate of change is defined as

$$\frac{dn}{dt} = f(n) \tag{9.44}$$

An equilibrium point is defined at a point, $\hat{n}$, where

$$\frac{dn}{dt} = f(\hat{n}) = 0 \tag{9.45}$$

Let there be a point close to the equilibrium point, distant from the latter by v; so

$$v = n - \hat{n} \tag{9.46}$$

307

where $n \simeq \hat{n}$ and so ν is small. So from (9.46)

$$\frac{dn}{dt} = \frac{d\nu}{dt} = f(\nu + \hat{n}) \tag{9.47}$$

The Taylor series expansion of (9.47) is

$$\frac{d\nu}{dt} = f(\hat{n}) + f'(\hat{n})\nu + O(\nu^2) \tag{9.48}$$

where $f'(\hat{n})$ is the derivative of $f(n)$ with respect to n, evaluated at $n = \hat{n}$. As ν is small, we need only consider the first two terms of the series, i.e.

$$\frac{dn}{dt} = \frac{d\nu}{dt} = f(\hat{n}) + f'(\hat{n})\nu \tag{9.49}$$

Since at equilibrium $f(\hat{n}) = 0$, we have

$$\frac{dn}{dt} = \frac{d\nu}{dt} = f'(\hat{n})\nu \tag{9.50}$$

i.e.

$$\frac{d\nu}{\nu} = f'(\hat{n}) \, dt$$

and so

$$\nu = a e\{f'(\hat{n})t\} \tag{9.51}$$

where a is the constant of integration. If $f(\hat{n})$ is negative, then $\nu \to 0$ as $t \to \infty$. In other words, the initial small displacement, ν, gets smaller with time; the equilibrium is stable. Conversely if $f'(\hat{n})$ is positive, then $\nu \to \infty$ as $t \to \infty$: the initial small displacement, ν, increases with time, and so the equilibrium is unstable. The case where $f'(\hat{n}) = 0$ is more complicated to deal with, and is not considered here.

Example 9.6   Investigate the equilibrium points of the logistic function

$$\frac{dn}{dt} = rn\left(1 - \frac{n}{K}\right) \tag{9.52}$$

where r and K are positive constants.

First, we find the equilibrium points, where $dn/dt = 0$. We have

$$r\hat{n}\left(1 - \frac{\hat{n}}{K}\right) = 0$$

308

So either $r\hat{n} = 0$, i.e. $\hat{n} = 0$, or $K - \hat{n} = 0$, i.e. $\hat{n} = K$. This accords with what we know about the curve of the integrated form of (9.52) – there are asymptotes at $n = 0$ and $n = K$ (the maximum value of n).

Now in the notation of (9.44)

$$f(n) = rn\left(1 - \frac{n}{K}\right)$$

and so

$$f'(n) = r - \frac{2r}{K}n$$

When $\hat{n} = 0$, $f'(\hat{n}) = r$, which is positive; and when $\hat{n} = K$, $f'(\hat{n}) = -r$, which is negative. So from (9.51) when $\hat{n} = 0$ the equilibrium is unstable; the curve of $n(t)$ moves away from $\hat{n} = 0$. Conversely, when $\hat{n} = K$ the equilibrium is stable, with the curve of $n(t)$ converging to its asymptote.

Two-variable system

Now we have two populations whose rates of change are defined by

$$\frac{dn_1}{dt} = f_1(n_1, n_2)$$

and

$$\frac{dn_2}{dt} = f_2(n_1, n_2)$$

$$(9.53)$$

An equilibrium point satisfies

$$\frac{dn_1}{dt} = f_1(\hat{n}_1, \hat{n}_2) = 0$$

and

$$\frac{dn_2}{dt} = f_2(\hat{n}_1, \hat{n}_2) = 0$$

$$(9.54)$$

simultaneously. The two-variable analogue of (9.46) is the equation pair

$$\left.\begin{array}{l} v_1 = n_1 - \hat{n}_1 \\ v_2 = n_2 - \hat{n}_2 \end{array}\right\}$$

$$(9.55)$$

So

$$\frac{dn_1}{dt} = \frac{dv_1}{dt} = f_1(v_1 + \hat{n}_1, v_2 + \hat{n}_2)$$

and

$$\frac{dn_2}{dt} = \frac{dv_2}{dt} = f_2(v_1 + \hat{n}_1, v_2 + \hat{n}_2)$$

$$(9.56)$$

309

DIFFERENTIAL EQUATIONS II

Next, we expand f_1 and f_2 in a Taylor series about the equilibrium point $(\hat{n}_1, \hat{n}_2)$, as far as the linear term only:

$$\left.\begin{aligned}
\frac{d\nu_1}{dt} &= f_1(\hat{n}_1, \hat{n}_2) + \left\{\frac{\partial f_1(\hat{n}_1, \hat{n}_2)}{\partial n_1}\right\}\nu_1 + \left\{\frac{\partial f_2(\hat{n}_1, \hat{n}_2)}{\partial n_2}\right\}\nu_2 \\
\frac{d\nu_2}{dt} &= f_2(\hat{n}_1, \hat{n}_2) + \left\{\frac{\partial f_2(\hat{n}_1, \hat{n}_2)}{\partial n_1}\right\}\nu_1 + \left\{\frac{\partial f_2(\hat{n}_1, \hat{n}_2)}{\partial n_2}\right\}\nu_2
\end{aligned}\right\} \quad (9.57)$$

Because $f_i(\hat{n}_1, \hat{n}_2) = 0$ for both $i = 1$ and 2, equation pair (9.57) is written as

$$\left.\begin{aligned}
\frac{d\nu_1}{dt} &= \left\{\frac{\partial f_1(\hat{n}_1, \hat{n}_2)}{\partial n_1}\right\}\nu_1 + \left\{\frac{\partial f_1(\hat{n}_1, \hat{n}_2)}{\partial n_2}\right\}\nu_2 \\
\frac{d\nu_2}{dt} &= \left\{\frac{\partial f_2(\hat{n}_1, \hat{n}_2)}{\partial n_1}\right\}\nu_1 + \left\{\frac{\partial f_2(\hat{n}_1, \hat{n}_2)}{\partial n_2}\right\}\nu_2
\end{aligned}\right\} \quad (9.58)$$

The pair of equations (9.58) is analogous to equation (9.50) for the one-variable system. Just as (9.50) may be solved to give ν as a function of t, so also may (9.58) be solved to give ν_1 and ν_2 as functions of t.

First, write

$$F_{ij} = \frac{\partial f_i(\hat{n}_1, \hat{n}_2)}{\partial n_j} \qquad i = 1, 2; \quad j = 1, 2$$

Then (9.58) appears more simply as

$$\frac{d\nu_1}{dt} = F_{11}\nu_1 + F_{12}\nu_2$$

$$\frac{d\nu_2}{dt} = F_{21}\nu_1 + F_{22}\nu_2$$

that is, a pair of linear equations. In matrix form, we have

$$\begin{bmatrix} F_{11} & F_{12} \\ F_{21} & F_{22} \end{bmatrix} \begin{bmatrix} \nu_1 \\ \nu_2 \end{bmatrix} = \begin{bmatrix} \dfrac{d\nu_1}{dt} \\ \dfrac{d\nu_2}{dt} \end{bmatrix}$$

and the matrix of partial derivatives is called the **Jacobian** (accent on the 'o'). Thus

$$\begin{bmatrix} \nu_1 \\ \nu_2 \end{bmatrix} = \begin{bmatrix} \dfrac{F_{22}}{\Delta} & -\dfrac{F_{12}}{\Delta} \\ -\dfrac{F_{21}}{\Delta} & \dfrac{F_{11}}{\Delta} \end{bmatrix} \begin{bmatrix} \dfrac{d\nu_1}{dt} \\ \dfrac{d\nu_2}{dt} \end{bmatrix} \quad (9.59)$$

310

where $\Delta = F_{11}F_{22} - F_{21}F_{12}$, the determinant of the Jacobian. On carrying out the matrix multiplication on the right-hand side of (9.59) and rearranging, we have

$$F_{22} \frac{d\nu_1}{dt} - F_{12} \frac{d\nu_2}{dt} - \Delta\nu_1 = 0 \tag{9.60}$$

$$-F_{21} \frac{d\nu_1}{dt} + F_{11} \frac{d\nu_2}{dt} - \Delta\nu_2 = 0 \tag{9.61}$$

a pair of linear first order differential equations with constant coefficients similar to, but simpler than, equations (9.1) and (9.2).

Next, by following through the steps from (9.1) and (9.2) to (9.4), we get

$$\nu_2 = \frac{1}{F_{12}\,\Delta} (F_{11}F_{22} - F_{12}F_{21}) \frac{d\nu_1}{dt} - \frac{F_{11}}{F_{12}} \nu_1$$

But, $F_{11}F_{22} - F_{12}F_{21} = \Delta$, and so

$$\nu_2 = \frac{1}{F_{12}} \frac{d\nu_1}{dt} - \frac{F_{11}}{F_{12}} \nu_1 \tag{9.62}$$

Differentiation of (9.62) with respect to t gives

$$\frac{d\nu_2}{dt} = \frac{1}{F_{12}} \frac{d^2\nu_1}{dt^2} - \frac{F_{11}}{F_{12}} \frac{d\nu_1}{dt} \tag{9.63}$$

Substituting (9.63) into (9.60) gives

$$F_{22} \frac{d\nu_1}{dt} - F_{12} \left\{ \frac{1}{F_{12}} \frac{d^2\nu_1}{dt^2} - \frac{F_{11}}{F_{12}} \frac{d\nu_1}{dt} \right\} - \Delta\nu_1 = 0$$

i.e.

$$\frac{d^2\nu_1}{dt^2} - (F_{11} + F_{22}) \frac{d\nu_1}{dt} + \Delta\nu_1 = 0$$

or

$$\frac{d^2\nu_1}{dt^2} - (F_{11} + F_{22}) \frac{d\nu_1}{dt} + (F_{11}F_{22} - F_{21}F_{12})\nu_1 = 0 \tag{9.64}$$

Equation (9.64) is a second-order linear homogeneous differential equation. Its auxiliary equation is

$$\lambda^2 - (F_{11} + F_{22})\lambda + (F_{11}F_{22} - F_{21}F_{12}) = 0$$

and the solution of (9.64) is

$$\nu_1 = c_1 \exp(\lambda_1 t) + c_2 \exp(\lambda_2 t) \tag{9.65}$$

Now consider the eigenvalues of the Jacobian. These are found by solving

311

DIFFERENTIAL EQUATIONS II

the equation

$$\begin{vmatrix} F_{11} - \lambda & F_{12} \\ F_{21} & F_{22} - \lambda \end{vmatrix} = 0$$

(see equation 1.8) which yields

$$(F_{11} - \lambda)(F_{22} - \lambda) - F_{21}F_{12} = 0$$

i.e.

$$F_{11}F_{22} - (F_{11} + F_{22})\lambda + \lambda^2 - F_{21}F_{12} = 0$$

or

$$\lambda^2 - (F_{11} + F_{22})\lambda + (F_{11}F_{22} - F_{21}F_{12}) = 0$$

Evidently the λs of (9.65) are the eigenvalues of the Jacobian; and since we obtain similar equations to (9.64) and (9.65) in v_2, the eigenvalues, λ_i, serve precisely the same function in the stability analysis of a two-variable system as does $f'(\hat{n})$ in equation (9.51) for the one-variable system. However, whereas in the one-variable system there were only two possible situations, $f'(\hat{n}) < 0$, $f'(\hat{n}) > 0$ ($f'(\hat{n}) = 0$ was not considered), the two-variable system provides many possibilities, and these are enumerated below.

(1) Real λ_i, $\lambda_1 < 0$, $\lambda_2 < 0$. Convergence occurs from both the n_1-direction and the n_2-direction; the equilibrium is termed a **stable node**. If $\lambda_1 = \lambda_2$ then the path of convergence from the displaced point to the stable node is linear; otherwise the trajectory is curvilinear.

(2) Real λ_i, $\lambda_1 > 0$, $\lambda_2 > 0$. This is the exact opposite of (1) above. Divergence occurs, and the equilibrium is called an **unstable node**.

(3) Real λ_i, $\lambda_1 < 0$, $\lambda_2 > 0$. Convergence occurs in the n_1-direction and divergence in the n_2-direction. It should be noted that convergence can only occur from a displacement exactly in the n_1-direction; divergence results from any other displacement after initial convergence, unless the displacement is exactly in the n_2-direction when there is no movement towards the equilibrium at all. The equilibrium is essentially unstable, and is called a **saddle point**. Similar remarks apply for $\lambda_1 > 0$ and $\lambda_2 < 0$.

The remaining three possibilities are for complex eigenvalues, when we can write (9.65) in the form

$$v_1 = c_1 \exp\{(a + ib)t\} + c_2 \exp\{(a - ib)t\}$$

(4) Complex λ_i, $a < 0$. The trajectory is a spiral which converges to the equilibrium point; the equilibrium point is called a **stable focus**.

(5) Complex λ_i, $a > 0$. The opposite of (4) above; the trajectory is a diverging spiral and the equilibrium point is termed an **unstable focus**.

(6) Complex λ_i, $a = 0$. In this case there is no convergence to, or divergence from, the equilibrium point. From a point displaced from the equilibrium the trajectory is a closed curve about the equilibrium point.

It is evident that in cases (4) to (6) the nature of the equilibrium is dependent solely on the real part of the complex eigenvalue.

Some more detail is given on the nature of the trajectories, under the different situations described above, by Roughgarden (1979).

Example 9.7 Investigate the equilibrium points of (a) the Lotka–Volterra, and (b) the Leslie and Gower models of host–parasite relationships.

(a) The Lotka–Volterra equations are quoted as equations (9.21) and (9.22), where all the constants are positive. At equilibrium, we have that

$$(a_1 - b_1 \hat{p})\hat{h} = 0$$

and

$$(- a_2 + b_2 \hat{h})\hat{p} = 0$$

giving

$$\hat{p} = a_1/b_1 \quad \text{and} \quad \hat{h} = a_2/b_2$$

The partial derivatives are

$$\frac{\partial f_1(h, p)}{\partial h} = a_1 - b_1 p$$

$$\frac{\partial f_1(h, p)}{\partial p} = - b_1 h$$

$$\frac{\partial f_2(h, p)}{\partial h} = b_2 p$$

$$\frac{\partial f_2(h, p)}{\partial p} = - a_2 + b_2 h$$

where $f_1(h, p)$ and $f_2(h, p)$ are, respectively, the right-hand sides of (9.21) and (9.22). Next we evaluate the partial derivatives at the equilibrium points; so

$$\frac{\partial f_1(\hat{h}, \hat{p})}{\partial h} = a_1 - b_1 \frac{a_1}{b_1} = 0$$

$$\frac{\partial f_1(\hat{h}, \hat{p})}{\partial p} = -b_1 \frac{a_2}{b_2} = -\frac{a_2 b_1}{b_2}$$

$$\frac{\partial f_2(\hat{h}, \hat{p})}{\partial h} = b_2 \frac{a_1}{b_1} = \frac{a_1 b_2}{b_1}$$

$$\frac{\partial f_2(\hat{h}, \hat{p})}{\partial p} = -a_2 + b_2 \frac{a_2}{b_2} = 0$$

So the Jacobian is

$$\mathbf{J} = \begin{bmatrix} 0 & -\dfrac{a_2 b_1}{b_2} \\ \dfrac{a_1 b_2}{b_1} & 0 \end{bmatrix}$$

The eigenvalues of this matrix are obtained from

$$(-\lambda)(-\lambda) - \left(\frac{a_1 b_2}{b_1}\right)\left(-\frac{a_2 b_1}{b_2}\right) = 0$$

i.e. from

$$\lambda^2 + a_1 a_2 = 0$$

Hence the eigenvalues are the conjugate pair

$$\lambda = \pm i\sqrt{(a_1 a_2)}$$

The real part of the eigenvalues is zero. Hence case (6) on page 313 applies, and the trajectories form closed curves around the equilibrium point, as shown in Figure 9.1.

(b)   The Leslie and Gower model is formulated in (9.24) and (9.25), again where all the constants are positive. At equilibrium, we have that

$$(a_1 - b_1 \hat{p})\hat{h} = 0$$

and

$$(a_2 - b_2 \hat{p}/\hat{h})\hat{p} = 0$$

giving

$$\hat{p} = \frac{a_1}{b_1} \quad \text{and} \quad \hat{h} = \frac{a_1 b_2}{a_2 b_1}$$

314

The partial derivatives are

$$\frac{\partial f_1(h, p)}{\partial h} = a_1 - b_1 p$$

$$\frac{\partial f_1(h, p)}{\partial p} = -b_1 h$$

$$\frac{\partial f_2(h, p)}{\partial h} = b_2 p^2 / h^2$$

$$\frac{\partial f_2(h, p)}{\partial p} = a_2 - 2b_2 p / h$$

By substituting the equilibrium values of h and p, we have the Jacobian

$$\mathbf{J} = \begin{bmatrix} 0 & -\dfrac{a_1 b_2}{a_2} \\[2ex] \dfrac{a_2^2}{b_2} & -a_2 \end{bmatrix}$$

The eigenvalues are obtained from

$$(-\lambda)(-a_2 - \lambda) - (a_2^2 / b_2)(-a_1 b_2 / a_2) = 0$$

which leads to

$$\lambda^2 + a_2 \lambda + a_1 a_2 = 0$$

Thus

$$\lambda = \frac{-a_2 \pm \sqrt{\{a_2(a_2 - 4a_1)\}}}{2}$$

If $a_2 < 4a_1$, the roots are complex. Since the real part $(-a_2/2)$ is negative, case (4) on page 312 applies and we have a stable focus with a trajectory spiralling in towards the equilibrium point. If $a_2 > 4a_1$, the roots are real. One of the roots is always negative, and both are if

$$a_2 > \sqrt{\{a_2(a_2 - 4a_1)\}}$$

If the latter applies, case (1) on page 312 results, giving a stable node; otherwise we have a saddle point and the equilibrium is unstable (case (3) on page 312).

Suggestions for further reading

Differential equations are introduced in most textbooks of calculus, and there are also many books devoted exclusively to the subject.

Of particular relevance to the biologist is **Jones D. S.** and **B. D. Sleeman** (1983). *Differential equations and mathematical biology*. London: Allen & Unwin. This is a textbook on differential equations in a biological setting.

The chapter on differential equations in **Grossman, S. I.** and **J. E. Turner** (1974). *Mathematics for the biological sciences*. New York: Collier Macmillan. This contains some interesting, but simple, biological applications.

Ecological applications of differential equations, mostly in the field of population dynamics, are given in **Maynard-Smith, J.** (1974). *Models in ecology*. Cambridge: Cambridge University Press, and **Pielou, E. C.** (1977). *Mathematical ecology*. New York: Wiley.

First-order and second-order equations, with physical applications, are presented in **Curle, N.** (1972). *Applied differential equations*. New York: Van Nostrand Reinhold.

Partial differential equations are thoroughly and clearly introduced by **Stephenson, G.** (1970). *An introduction to partial differential equations for science students*, 2nd edn. London: Longman.

Stability analysis is outlined in Appendix 3 of **Roughgarden, J.** (1979). *Theory of population genetics and evolutionary ecology: an introduction*. West Drayton: Collier Macmillan. The same appendix also contains an interesting review of linear algebra, which could profitably be read after Chapter 1 of this book.

A useful book on mathematical models in biology is **Finkelstein, L.** and **E. R. Carson** (1979). *Mathematical modeling of dynamic biological systems*. Medical Computing Series, Vol. III. Forest Grove, Oreg.: Research Studies Press.

Answers to exercises

1.1 (a) 32   (b) -6

1.2 (a) -160 (b) -264

1.3 (a) $\begin{bmatrix} 6 & 4 & 6 \\ 24 & 24 & 16 \\ 5 & 13 & -3 \end{bmatrix}$ (b) $\begin{vmatrix} -6 & 3 & 5 \\ 12 & 4 & 4 \\ 14 & 23 & 29 \end{vmatrix}$

(c) 20   (d) 4

(e) The determinant of the product matrix **AB** (in (a)) is 80.

(f) The determinant of the product matrix **BA** (in (b)) is 80.

This shows that $|AB| = |BA| = |A\,\|\,B| = |B\,\|\,A|$.

1.4   adj $\mathbf{A} = \begin{bmatrix} 2 & -3 & 8 \\ 14 & -1 & -4 \\ -10 & 5 & 0 \end{bmatrix}$ adj $\mathbf{B} = \begin{bmatrix} -17 & -14 & 5 \\ 10 & 8 & -2 \\ 11 & 10 & -3 \end{bmatrix}$

$\mathbf{A}^{-1} = \begin{bmatrix} \frac{1}{10} & -\frac{3}{20} & \frac{2}{5} \\ \frac{7}{10} & -\frac{1}{20} & -\frac{1}{5} \\ -\frac{1}{2} & \frac{1}{4} & 0 \end{bmatrix}$ $\mathbf{B}^{-1} = \begin{bmatrix} -4\frac{1}{4} & -3\frac{1}{2} & 1\frac{1}{4} \\ 2\frac{1}{2} & 2 & -\frac{1}{2} \\ 2\frac{3}{4} & 2\frac{1}{2} & -\frac{3}{4} \end{bmatrix}$

1.5 $|A^{-1}| = 0.05$ $|B^{-1}| = 0.25$

This shows that the determinant of an inverse matrix is equal to the inverse of the determinant of the matrix itself, i.e. $|A^{-1}| = 1/|A| = |A|^{-1}$.

1.6 (a) $x_1 = 26$ $x_2 = 5$ $x_3 = -11.4$

(b) $x_1 = -208$ $x_2 = -141$ $x_3 = -9$ $x_4 = 13$

(c) $x_1 = 0.4x_3 + 43.2$ $x_2 = -1.4x_3 - 19.2$

Put $x_3 = \lambda$, then

$$\begin{bmatrix} x_1 \\ x_2 \\ x_3 \end{bmatrix} = \lambda \begin{bmatrix} 0.4 \\ -1.4 \\ 1.0 \end{bmatrix} + \begin{bmatrix} 43.2 \\ -19.2 \\ 0 \end{bmatrix}$$

(d) $\begin{bmatrix} x_1 \\ x_2 \\ x_3 \end{bmatrix} = \lambda \begin{bmatrix} 5 \\ -3 \\ 2 \end{bmatrix} + \frac{1}{2} \begin{bmatrix} -1 \\ 1 \\ 0 \end{bmatrix}$

(e) $\begin{bmatrix} x_1 \\ x_2 \\ x_3 \end{bmatrix} = \lambda \begin{bmatrix} -14 \\ 11 \\ 16 \end{bmatrix}$

ANSWERS TO EXERCISES

(f) Equations are incompatible, hence no solution.

(g) $\begin{bmatrix} x_1 \\ x_2 \\ x_3 \end{bmatrix} = \lambda \begin{bmatrix} 1 \\ -2 \\ 1 \end{bmatrix} + \begin{bmatrix} 1 \\ 1 \\ 0 \end{bmatrix}$

1.7 (a) Eigenvalue = 14 with eigenvector

$$\begin{bmatrix} 1 \\ 3 \end{bmatrix}$$

Eigenvalue = 1 with eigenvector

$$\begin{bmatrix} 4 \\ -1 \end{bmatrix}$$

Diagonal form

$$\begin{bmatrix} 14 & 0 \\ 0 & 1 \end{bmatrix}$$

(b) Eigenvalue = 2 with eigenvector

$$\begin{bmatrix} 1 \\ -1 \\ 0 \end{bmatrix}$$

Eigenvalue = 3 with eigenvector

$$\begin{bmatrix} 1 \\ 0 \\ -1 \end{bmatrix}$$

Eigenvalue = 4 with eigenvector

$$\begin{bmatrix} 1 \\ -1 \\ -1 \end{bmatrix}$$

Diagonal form

$$\begin{bmatrix} 2 & 0 & 0 \\ 0 & 3 & 0 \\ 0 & 0 & 4 \end{bmatrix}$$

1.8 (a) 3, full rank (b) 4, full rank (c) 1
(d) 2 (e) 2 (f) 2
(g) 2

1.9 (c) 2, −7, 5 (d) 5, −3, 2 (e) −14, 11, 16
(f) 1, −2, 1 (g) 1, −2, 1

3.1 (a) The substitution yields $\int dz/(z^2 - 1)$. This is then resolved into partial fractions, making use of the fact that $(z-1)(z+1) = z^2 - 1$. The final answer is $\frac{1}{2}[\log_e\{\sqrt{(1+x^2)} - 1\} - \log_e\{\sqrt{(1+x^2)} + 1\}] + c$.

318

(b) The substitution gives $\frac{1}{4}\int (z-1)\,dz/z^3$, and the final answer is
$\frac{1}{4}[1/\{2(2x+1)^2\} - 1/(2x+1)] + c$.

(c) $\frac{1}{2}\log_e(1+x^2) + c$

3.2 (a) $e^x(x-1) + c$

(b) The formula for integration by parts gives

$$x^2 e^x - 2 \int xe^x \, dx + c.$$

Using (a) above, we finally get $e^x(x^2 - 2x + 2) + c$.

(c) $\log_e x$ must be represented by $\psi(x)$ in the formula for integration by parts. The answer is $\frac{1}{2}x^2(\log_e x - \frac{1}{2}) + c$.

(d) $(-1/x)(\log_e x + 1) + c$.

3.3 (a) Using the result of 3.2(a), we obtain the answer 1.

(b) The integral is $\tan^{-1}x$. So $[\tan^{-1}x]_{-\infty}^{\infty} = \pi$.

(c) The integral evaluates to

$$2\left[\lim_{x\to\infty} (\sqrt{x}) - 1\right],$$

which cannot be evaluated.

(d) $\frac{1}{2}$

3.4 We require that

$$0.3 \int_0^l e^{-0.3x} \, dx = 0.025.$$

Evaluating the integral finally yields $1 - e^{-0.3l} = 0.025$, from which $l = 0.0844$. Similarly, we require that

$$0.3 \int_0^u e^{-0.3x} \, dx = 0.975,$$

which leads to $u = 12.2963$. The probability of finding an observation in the range $0.0844 \leqslant x \leqslant 12.2963$ is 0.95.

3.5 (a) $\frac{1}{2}\sqrt{\pi}$ (b) $\frac{3}{4}\sqrt{\pi}$ (c) $\frac{15}{8}\sqrt{\pi}$ (d) $\frac{105}{16}\sqrt{\pi}$

3.6 (a) $\frac{1}{2}\pi$ (b) $\frac{1}{8}\pi$ (c) $\frac{3}{48}\pi$ (d) $\frac{9}{384}\pi$

3.8 $a = 1.5$ $V = \frac{3}{7}\{1/(1-K)\}^{4/3}\pi d^2 H/4$

3.9 1

3.10 $\frac{3}{48}\pi a^4$

4.1 (a) 1.0472 (b) 1.7453 (c) 0.2618 rad

4.2 (a) $5°44'$ (b) $154°42'$ (c) $80°13'$

4.3 (a) $2x(x\cos 2x + \sin 2x)$
(b) $(\tan x - x\sec^2 x)/\tan x$
(c) $3\sin(2 - 3x)$

319

4.4 $dy/dx = 10e^{0.2 \sin x} \cos x$. Equating to zero yields the result $\cos x = 0$; so $x = \pi/2$ or $3\pi/2$ in the required range. Substituting these values successively into the original equation gives $y = 61.07$ or 40.94, respectively. $d^2y/dx^2 = 2e^{0.2 \sin x}(\cos^2 x - \sin^2 x)$. When $x = \pi/2$, d^2y/dx^2 is negative; hence the point $(\pi/2, 61.07)$ is a maximum. When $x = 3\pi/2$, d^2y/dx^2 is positive; hence the point $(3\pi/2, 40.94)$ is a minimum.

4.5 (a) $-\frac{1}{4}\cos 4x + c$ (b) $-\frac{1}{2}e^{-x} + c$ (c) $\frac{1}{2}\sin^2 x + c$

4.6 (a) $x \sin x + \cos x + c$ (b) $\frac{1}{2}e^x(\sin x - \cos x) + c$

4.7 (a) $2/\sqrt{(1 - 4x^2)}$ (b) $1/\sqrt{\{x(2 - x)\}}$ (c) $3/(1 + 9x^2)$

4.8 (a) $\sin^{-1}\frac{1}{2}x$ (b) $\frac{1}{2}\sin^{-1}(3x/2)$

4.10 $(\sinh \pi)/\pi\left[1 + 2\sum_{r=1}^{\infty}\{(-1)^r/(r^2 + 1)\}\{\cos rx - r\sin rx\}\right]$

5.1 (a) $1 + x + x^2 + x^3 + \ldots$
(b) $2 + \frac{1}{4}x^2 - \frac{1}{64}x^4 + \frac{1}{512}x^6 - \ldots$
(c) $4 - \frac{1}{3}x^3 - \frac{1}{144}x^6 - \frac{1}{2592}x^9 - \ldots$

5.3 (a) 0.7661   (b) 0.9397   (c) 0.6932   (d) -0.2231

5.4 (a) $r = 4.4721$ $\theta = -0.4637$
(b) $r = 5.2915$ $\theta = 0.3335$
(c) $r = 4.4721$ $\theta = 0.4637$
(d) $r = 5.2915$ $\theta = -0.3335$

5.5 (a) $-1 - 6.7321i$ (b) 8   (c) $-4i$ (d) -10
(e) $-3.4641i$ (f) $-16.5359 + 16.9282i$ (g) 20
(h) 28   (i) $1.1732 + 0.1536i$ (j) $0.6 - 0.8i$
(k) $0.7857 + 0.6186i$

5.8 (a) $3/(x - 2) - 2/(2x + 1)$
(b) $3/(x + 2) - 5/(x + 2)^2$
(c) $(2x - 1)/(x^2 + 2) - 2/(x + 3)$

5.9 Discriminant is $(-3)^3 + 27(3)^2 = -27 + 243 = 216$, which is positive. Hence the equation has only one real root. By sketching the curve of the function, the root is found to be near -2. Putting $x_0 = -2$, then $x_1 = -2.1$, $x_2 = -2.103\ 836$, $x_3 = -2.103\ 803$. Hence the real root of the equation is -2.1038 correct to four decimal places.

6.1 $\partial z/\partial x = -ke^{-kx}\cos ay$
$\partial z/\partial y = -ae^{-kx}\sin ay$
$\partial^2 z/\partial x^2 = k^2 e^{-kx}\cos ay$
$\partial^2 z/\partial y^2 = -a^2 e^{-kx}\cos ay$
$\partial^2 z/\partial x\partial y = ake^{-kx}\sin ay$

6.2 $\partial z/\partial x = 3x^2 + y$ $\partial z/\partial y = x + 2y$
Equating to zero and solving the pair of equations simultaneously gives $(0, 0, 0)$ and $(\frac{1}{6}, -\frac{1}{12}, -\frac{1}{432})$ as stationary points.

$$\partial^2 z/\partial x^2 = 6x \qquad \partial^2 z/\partial y^2 = 2 \qquad \partial^2 z/\partial x\,\partial y = 1$$

When $x = 0$, $\partial^2 z/\partial x^2 = 0$, and so this stationary point is a saddle point. When $x = \frac{1}{6}$, $\partial^2 z/\partial x^2 = 1$, showing this to be a minimum point.

7.1 Form the quantity $S^2 = \sum \{y - \bar{y} - b(x - \bar{x})\}^2$;

then $\partial S^2/\partial \bar{y} = -2\sum \{y - \bar{y} - b(x - \bar{x})\}$

and $\partial S^2/\partial b = -2\sum \{y - \bar{y} - b(x - \bar{x})\}(x - \bar{x})$.

Equating the partial derivatives to zero and rearranging gives the pair of linear equations

$$\bar{y}n \qquad + b\sum(x - \bar{x}) = \sum y$$
$$\bar{y}\sum(x - \bar{x}) + b\sum(x - \bar{x})^2 = \sum(x - \bar{x})y$$

Since $\sum(x - \bar{x}) = 0$, the first equation gives $\bar{y} = (1/n)\sum y$. Also, since it can be shown that $\sum(x - \bar{x})y = \sum(x - \bar{x})(y - \bar{y})$, the second equation gives $b = \{\sum(x - \bar{x})(y - \bar{y})\}/\{\sum(x - \bar{x})^2\}$.

7.2 Form the quantity $S^2 = \sum(y - a\cos x - b\sin x - c)^2$; differentiate successively with respect to a, b, and c, and equate to zero. On rearranging, we have

$$a\sum \cos^2 x + b\sum \sin x \cos x + c\sum \cos x = \sum y\cos x$$
$$a\sum \cos x \sin x + b\sum \sin^2 x + c\sum \sin x = \sum y\sin x$$
$$a\sum \cos x + b\sum \sin x + cn = \sum y$$

Although it is possible to obtain explicit formulae for a, b, and c from these equations, there would be no virtue in doing this since the set of equations can be easily solved numerically for specific data sets.

8.1 (a) Applying the integrating factor gives $y = e^{-2x}\int xe^{2x}\,dx + c$. Integrating by parts gives the final solution: $y = \frac{1}{2}x + ce^{-2x} - \frac{1}{4}$.

(b) Applying the integrating factor gives $y = 2(\operatorname{cosec} x)(\int \cos 2x \sin x\,dx + c)$. Integrating by parts twice gives the final solution: $y = \frac{2}{3}(\operatorname{cosec} x)(\cos 2x \cos x + 2\sin 2x \sin x + c)$.

(c) The equation first has to be divided throughout by $\cos x$, giving $dy/dx + y\tan x = \sec x$. The solution is $y = \sin x + c\cos x$.

8.2 (a) $y = 2e^{-x}(A\cos 2x - B\sin 2x) + \frac{1}{5}x^2 - \frac{4}{25}x + \frac{23}{125}$

(b) $y = (c_1 + \frac{1}{3}x)e^{-x} + (c_2 - 2x)e^{-2x}$

(c) $y = 2A\cos 2x + (\frac{1}{4}x - 2B)\sin 2x$

8.3 $y = (c_1 - 1)e^{4x} + (c_2 + 1)e^{3x}$

8.4 $y = Ax^3 + Bx^2 - 2x^2\cos x - x^3\sin x$

Index

Bold numbers indicate a main section, which may extend over more than one page, and/or a definition. Italicised numbers refer to pages containing the subject in a text figure or table.

DATE DUE

JUN 2 9 1990			

DEMCO 38-297